PUBLIC RESPONSE
TO ALERTS AND WARNINGS
USING SOCIAL MEDIA

REPORT OF A WORKSHOP ON CURRENT
KNOWLEDGE AND RESEARCH GAPS

Committee on Public Response to Alerts and Warnings Using Social Media:
Current Knowledge and Research Gaps

Computer Science and Telecommunications Board

Division on Engineering and Physical Sciences

NATIONAL RESEARCH COUNCIL
OF THE NATIONAL ACADEMIES

THE NATIONAL ACADEMIES PRESS
Washington, D.C.
www.nap.edu

THE NATIONAL ACADEMIES PRESS   500 Fifth Street, NW   Washington, DC 20001

NOTICE: The project that is the subject of this report was approved by the Governing Board of the National Research Council, whose members are drawn from the councils of the National Academy of Sciences, the National Academy of Engineering, and the Institute of Medicine. The members of the committee responsible for the report were chosen for their special competences and with regard for appropriate balance.

Support for this project was provided by the Department of Homeland Security with assistance from the National Science Foundation under award number IIS-1118399. Any opinions, findings, or conclusions expressed in this publication are those of the authors and do not necessarily reflect the views of the organizations that provided support for the project.

International Standard Book Number-13:   978-0-309-29033-3
International Standard Book Number-10:   0-309-29033-3

This report is available from:

Computer Science and Telecommunications Board
National Research Council
500 Fifth Street, NW
Washington, DC 20001

Additional copies of this report are available from:

The National Academies Press
500 Fifth Street, NW, Keck 360
Washington, DC 20001
(800) 624-6242 or (202) 334-3313
http://www.nap.edu

Copyright 2013 by the National Academy of Sciences. All rights reserved.

Printed in the United States of America

THE NATIONAL ACADEMIES
Advisers to the Nation on Science, Engineering, and Medicine

The **National Academy of Sciences** is a private, nonprofit, self-perpetuating society of distinguished scholars engaged in scientific and engineering research, dedicated to the furtherance of science and technology and to their use for the general welfare. Upon the authority of the charter granted to it by the Congress in 1863, the Academy has a mandate that requires it to advise the federal government on scientific and technical matters. Dr. Ralph J. Cicerone is president of the National Academy of Sciences.

The **National Academy of Engineering** was established in 1964, under the charter of the National Academy of Sciences, as a parallel organization of outstanding engineers. It is autonomous in its administration and in the selection of its members, sharing with the National Academy of Sciences the responsibility for advising the federal government. The National Academy of Engineering also sponsors engineering programs aimed at meeting national needs, encourages education and research, and recognizes the superior achievements of engineers. Dr. Charles M. Vest is president of the National Academy of Engineering.

The **Institute of Medicine** was established in 1970 by the National Academy of Sciences to secure the services of eminent members of appropriate professions in the examination of policy matters pertaining to the health of the public. The Institute acts under the responsibility given to the National Academy of Sciences by its congressional charter to be an adviser to the federal government and, upon its own initiative, to identify issues of medical care, research, and education. Dr. Harvey V. Fineberg is president of the Institute of Medicine.

The **National Research Council** was organized by the National Academy of Sciences in 1916 to associate the broad community of science and technology with the Academy's purposes of furthering knowledge and advising the federal government. Functioning in accordance with general policies determined by the Academy, the Council has become the principal operating agency of both the National Academy of Sciences and the National Academy of Engineering in providing services to the government, the public, and the scientific and engineering communities. The Council is administered jointly by both Academies and the Institute of Medicine. Dr. Ralph J. Cicerone and Dr. Charles M. Vest are chair and vice chair, respectively, of the National Research Council.

www.national-academies.org

**COMMITTEE ON PUBLIC RESPONSE TO ALERTS
AND WARNINGS USING SOCIAL MEDIA: CURRENT
KNOWLEDGE AND RESEARCH GAPS**

ROBERT E. KRAUT, Carnegie Mellon University, *Chair*
ALESSANDRO ACQUISTI, Carnegie Mellon University
JON M. KLEINBERG, Cornell University
LESLIE LUKE, San Diego County Office of Emergency Services
RICHARD G. MUTH, State of Maryland Emergency Management Agency
LEYSIA PALEN, University of Colorado, Boulder
TIMOTHY L. SELLNOW, University of Kentucky
MICHELE WOOD, California State University, Fullerton

Staff

JON EISENBERG, Director, Computer Science and Telecommunications Board
VIRGINIA BACON TALATI, Associate Program Officer
ERIC WHITAKER, Senior Program Assistant

COMPUTER SCIENCE AND TELECOMMUNICATIONS BOARD

ROBERT F. SPROULL, Sun Labs, *Chair*
PRITHVIRAJ BANERJEE, Hewlett Packard Company
STEVEN M. BELLOVIN, Columbia University
JACK L. GOLDSMITH III, Harvard Law School
SEYMOUR E. GOODMAN, Georgia Institute of Technology
JON M. KLEINBERG, Cornell University
ROBERT E. KRAUT, Carnegie Mellon University
SUSAN LANDAU, Radcliffe Institute for Advanced Study
PETER LEE, Microsoft Corporation
DAVID E. LIDDLE, US Venture Partners
DAVID E. SHAW, D.E. Shaw Research
ALFRED Z. SPECTOR, Google, Inc.
JOHN STANKOVIC, University of Virginia
JOHN A. SWAINSON, Dell, Inc.
PETER SZOLOVITS, Massachusetts Institute of Technology
PETER J. WEINBERGER, Google, Inc.
ERNEST J. WILSON, University of Southern California
KATHERINE YELICK, University of California, Berkeley

Staff

JON EISENBERG, Director
VIRGINIA BACON TALATI, Associate Program Officer
SHENAE BRADLEY, Senior Program Assistant
RENEE HAWKINS, Financial and Administrative Manager
HERBERT S. LIN, Chief Scientist
LYNETTE I. MILLETT, Associate Director
ERIC WHITAKER, Senior Program Assistant
ENITA A. WILLIAMS, Associate Program Officer

For more information on CSTB, see its website at http://www.cstb.org, write to CSTB, National Research Council, 500 Fifth Street, NW, Washington, DC 20001, call (202) 334-2605, or e-mail the CSTB at cstb@nas.edu.

Preface

Following an earlier workshop organized by a separate National Research Council (NRC) committee that explored the public response to alerts and warnings delivered to mobile devices,[1] the Department of Homeland Security's (DHS's) Science and Technology Directorate asked the NRC's Computer Science and Telecommunications Board to convene a workshop on the role of social media in disaster response. Held February 28 and 29, 2012, the workshop was organized by the Committee on Public Response to Alerts and Warnings Using Social Media: Current Knowledge and Research Gaps. The resulting report points to potential topics for future research and possible areas for future research investment by DHS and others and describes challenges facing disaster managers seeking to incorporate social media into regular practice.

One of the first workshops convened to look systematically at the use of social media for alerts and warnings, the event brought together social science researchers, technologists, emergency management professionals, and other experts on how the public and emergency managers use social media in disasters. The workshop explored (1) what is known about how the public responds to alerts and warnings; (2) the implications of what is known about such public responses for the use of social media to provide alerts and warnings to the public; and (3) approaches to enhancing the

---

[1] National Research Council. *Public Response to Alerts and Warnings on Mobile Devices: Summary of a Workshop on Current Knowledge and Research Gaps*. The National Academies Press, Washington, D.C., 2011.

> **BOX P.1**
> **Statement of Task**
>
> An ad hoc committee will oversee development and facilitation of a workshop that convenes experts from across the nation in the fields of alerts and warnings, social media, and privacy. The workshop will examine the use of and public response to social media for alerts, identifying past and current research and future research needs. It will also explore potential privacy implications of issuing alerts and warnings via social media. The workshop will use a mix of individual presentations, panels, breakout discussions, and question-and-answer sessions to develop an understanding of the relevant research communities, research already completed, ongoing research, and future research needs. Key stakeholders will be identified and invited to participate. An unedited (verbatim) transcript of the event will be prepared. A report summarizing what transpired at the workshop will be prepared.

situational awareness of emergency managers. It also considered how officials monitor social media and the privacy considerations that result. The complete statement of task for the workshop is provided in Box P.1.

This report summarizes presentations made by invited speakers, other remarks by workshop participants, and discussion during parallel breakout sessions. In keeping with the workshop's purpose of exploring an emerging topic, this summary does not contain findings or recommendations. Nor, in keeping with NRC guidelines for workshop reports, does it necessarily reflect consensus views of the workshop participants or the responsible committee. In addition, these summaries should not be taken as remarks made solely by the scheduled session speakers, because the discussions included remarks offered by others in attendance, and the summaries of the workshop sessions provided in the chapters of this report are a digest both of the presentations and of the subsequent discussion.

Chapter 1 provides a brief overview of background information on the alerting process and public response as well as current understanding of social media use. Chapters 2 through 5 provide integrated summaries of the session presentations and the discussion that followed, organized by topic. Chapter 6 summarizes the research questions identified during the breakout sessions and subsequent plenary discussion. Appendix A presents the workshop agenda, and speaker biosketches are provided in Appendix B. Appendix C provides biosketches of the committee and the staff.

>
> Robert E. Kraut, *Chair*
> Committee on Public Response to
> Alerts and Warnings Using Social Media:
> Current Knowledge and Research Gaps

Acknowledgment of Reviewers

This report has been reviewed in draft form by individuals chosen for their diverse perspectives and technical expertise, in accordance with procedures approved by the National Research Council's Report Review Committee. The purpose of this independent review is to provide candid and critical comments that will assist the institution in making its published report as sound as possible and to ensure that the report meets institutional standards for objectivity, evidence, and responsiveness to the study charge. The review comments and draft manuscript remain confidential to protect the integrity of the deliberative process. We wish to thank the following individuals for their review of this report:

Robert Dudgeon, San Francisco Department of Emergency Management
Prabhakar Raghavan, Google
Ellis Stanley, Independent Consultant, Roswell, Georgia
Clarence L. Wardell III, CNA Safety
Duncan Watts, Microsoft Research

Although the reviewers listed above have provided many constructive comments and suggestions, they were not asked to endorse the material presented, nor did they see the final draft of the report before its release. The review of this report was overseen by Ruzena Bajcsy, University of California, Berkeley. Appointed by the National Research Council, she was responsible for making certain that an independent examination of this report was carried out in accordance with institutional procedures and that all review comments were carefully considered. Responsibility for the final content of this report rests entirely with the authoring committee and the institution.

Contents

1   FUNDAMENTALS OF ALERTS, WARNINGS, AND SOCIAL MEDIA   1
Current Knowledge About Public Response to Alerts and Warnings, 2
Social Media Use by the Public During Disasters, 4
Barriers to Incorporating Social Media into Emergency Management, 5
Technology Development for the Use of Social Media in Emergency Management, 6
Technologies for Developing Situational Awareness from Social Media, 9
Observations of Workshop Participants, 10

2   CURRENT USES OF SOCIAL MEDIA IN EMERGENCIES   12
Use of Social Media by the Los Angeles Fire Department, 12
Use of Social Media by WCNC, Charlotte, to Provide Weather Information, 14
Using Social Media for Earthquake Detection and Alerting, 16
Using Social Media to Assess Communication Needs and Disseminate Information During a Health Emergency, 17
Observations of Workshop Participants, 20

3 DYNAMICS OF SOCIAL MEDIA   22
   Studying Twitter Use to Understand How People
      Communicate, 22
   Problem Solving with Social Media, 25
   Standby Task Force: Volunteer Networks During Disasters, 28
   Observations of Workshop Participants, 32

4 CREDIBILITY, AUTHENTICITY, AND REPUTATION   34
   Reputation Systems, 34
   Encouraging Self-Correction, 36
   Computational Claim Verification, 36
   Applying the "Citizen Science" Model to Disaster
      Management, 37
   Observations of Workshop Participants, 40

5 PRIVACY AND LEGAL CHALLENGES WITH THE USE OF
   SOCIAL MEDIA   41
   Legal and Policy Perspectives on Privacy and on
      Government Monitoring of Social Media, 41
   Privacy Protection in the Context of Programs for Citizen
      Reporting of Threats, 44
   Legal Perspective on First-Responder Responsibilities, 45
   Observations of Workshop Participants, 47

6 RESEARCH GAPS AND IMPLEMENTATION CHALLENGES   49
   Message Content and Dissemination, 49
   Trust and Credibility, 51
   Privacy, 52
   Volunteers, 52
   Technology Diffusion, 53
   Emergency Management Practice, 54

APPENDIXES

A Workshop Agenda   57
B Biosketches of Workshop Speakers   63
C Biosketches of Committee and Staff Members   73

1

Fundamentals of Alerts, Warnings, and Social Media

The Warning, Alert, and Response Network (WARN) Act of 2006 called for the creation of a national all-hazards alerting system that would use multiple technologies to better reach affected populations. Since its passage, there has been an increasing interest in exploring the use of social media to provide alerts and warnings. *Social media* is a loosely defined term that refers to a set of Internet-based tools that support social interaction through many-to-many communications. The term encompasses a variety of technologies including weblogs, microbloging and mashup tools, and online social networks, examples of which include Twitter, Google Maps, Facebook, and Flickr. It also includes tools like Ushahidi that were purpose-built for use in crises. Social media are being used in all sectors of society to support communication, collaboration, and information collection and dissemination. Social media have proven useful in a crisis both to officials seeking to deliver alerts, warnings, and other information to the public and to citizens communicating with officials and each other.

This report presents a summary of a February 2012 workshop organized by the National Research Council's (NRC's) Committee on Public Response to Alerts and Warnings on Using Social Media: Current Knowledge and Research Gaps. The first session of the workshop provided an overview of alerts (an *alert* indicates that something significant has happened or may happen), warnings (a *warning* typically follows an alert and provides more detailed information indicating what protective action should be taken),[1] the use of social media for delivering alerts and warnings, and other applications of social media in disaster management. Dennis Mileti, University of Colorado, Boulder, described what is known about how the public responds to alerts and warnings. Kristiana Almeida, American Red Cross (ARC), described how the ARC uses social media during disasters and provided results of ARC research on social media use. Edward Hopkins, Maryland Emergency Management Agency, discussed barriers to the use of social media by emergency managers. Emre Gunduzhan, Johns Hopkins University Applied Physics Laboratory, described current and emerging technologies for disseminating alerts and warnings and enhancing situational awareness using social media.

CURRENT KNOWLEDGE ABOUT PUBLIC RESPONSE TO ALERTS AND WARNINGS

More than 60 years of interdisciplinary research on disaster response has yielded many insights about how people respond to information indicating that they are at risk and under what circumstances they are most likely to take appropriate protective action. Much of this knowledge has been captured in the "Annotated Bibliography for Public Risk Communication on Warnings for Public Protective Action Response and Public Education" that lists more than 350 publications.[2] This body of research covers natural disasters such as Hurricane Camille and the Mount St. Helens eruption; terrorist attacks such as those on the World Trade Center in 1993 and 2001; hazardous material spills such as those that occurred during the 1979 Mississauga, Ontario, train derailment and the 1987 Nanticoke, Pennsylvania, factory fire; building fires such as those at the MGM Grand Hotel in Las Vegas in 1980 and at Chicago's Cook County Hospital; and technological accidents such as the 1979 incident at the Three Mile Island nuclear power station in Pennsylvania. Mileti outlined some of the key results from research on how the public responds to alerts and warnings:

---

[1] The difference between alerts and warnings can be unclear because a warning can also serve as an alert, and an alert may be accompanied by some information about protective measures. Technology has further eroded the distinction. For example, on mobile devices, the Commercial Mobile Alert service will simultaneously deliver both a distinctive tone (the alert) and a brief message with additional information (a warning). Similarly, sirens have evolved to provide both a siren sound and a spoken message.

[2] The extensive "Annotated Bibliography for Public Risk Communication on Warnings for Public Protective Actions Response and Public Education" was compiled by Dennis Mileti, Rachel Bandy, Linda B. Bourque, Aaron Johnson, Megumi Kano, Lori Peck, Jeannette Sutton, and Michele Wood and is available at www.colorado.edu/hazards/publications/informer/infrmr2/pubhazbibann.pdf.

- *Research using hypothetical scenarios does not faithfully represent public response.* Studies on how people would respond to hypothetical events do not generally predict how the public is likely to behave in response to an actual alert or warning—or during a real event. By contrast, although they may be harder or more expensive to conduct, studies of actual events yield much better insights about the situational determinants of behavior during emergencies.
- *Education about the warning system is needed before an event.* Public education on warning systems is an important complement to education about how to prepare for disasters. For example, it is important that people know that a certain television or cell phone tone designates an alert or warning, and where additional sources of authoritative information can be found. Furthermore, it is important to remember that the majority of people cannot remember what a given siren means, what a color code may represent, or even the difference between the watches and warnings issued by the National Weather Service. Comprehensive education programs teach the public that a hazard exists, inform them about the alerting/warning systems in place in their communities, and outline what protective actions they might be asked to take. Essentially, these programs prime the public by removing surprises and reducing confusion in a future warning event.
- *Alerting needs to attract attention.* A primary goal of alerting is to attract the affected population's attention so that one can then provide information. The most effective alerts are incredibly intrusive, able to be noticed amidst the cacophony of daily life, and, for those who are asleep, literally loud enough to wake them. The more channels, or mechanisms used to disseminate alerts and warnings, the greater the chance an individual will receive a message. Warnings should also be repeated; repetition commands attention, fosters confirmation, and prompts protective action.
- *People seek social confirmation of warnings before taking protective action.* Before acting in response to a warning, people generally seek confirmation from others. The resulting process is known as milling, in which individuals interact with others to confirm information and develop a view about the risks they face at that moment and their possible responses. Milling creates a lag between the time a warning is received and the time protective action is taken. Yet few approaches to formulating and delivering warnings focus on shortening this milling time. Indeed, one important lesson from past research is that without careful attention to the process of milling, the introduction of new warning systems may increase rather than decrease individuals' delay in response. Social media provide a new way for these interactions among individuals to occur. Although many first responders believe that social media have given

them less control over the warning process, informal dissemination of messages has always played an important role in the warning process. Indeed, one might conjecture that the inherently social nature of social media might help reduce milling time, but whether this is true and under what conditions is an open research question.

- *Messages should contain information that is important to the population.* The key content of messages is what (the protective actions that should be taken); when (by what time the protective actions should be taken); where (the geographic area that will be affected); why (the risks and how protective action would reduce their impact); and who (the individuals or entities providing the information). In addition to content, message style also matters. Messages need to be clear, simply worded; specific (i.e., precise and non-ambiguous); accurate (i.e., free from errors that can create confusion); certain (i.e., authoritative and confident); and consistent. When changes in instructions are required, the reasons should be explained.
- *Responders should consider the demographics of affected populations when preparing warning messages.* Status differences (e.g., gender, sex, age, ethnicity, socioeconomics, and family relationships), an individual's or a community's past experiences, and other environmental and social factors all can affect how a warning will be interpreted.
- *Access for those with disabilities must be considered when developing alert or warning systems.* Individuals with impaired hearing or impaired vision or those with limited abilities may face challenges in receiving particular alerts or warnings. Furthermore, individuals with disabilities may encounter unique challenges in taking protective action or there may be important unique factors in how they may interpret a warning.
- *Alerting and warning is a process, not a single act.* The communications process includes issuing a warning, monitoring the public's response to the warning, listening for incorrect information the public may be receiving, and rewarning based on observations of what the public is doing or not doing. Given the mix of official and unofficial information available to the public, it is almost inevitable that people will be exposed to some incorrect information, which can lead to inconsistencies that can delay protective actions. Such misinformation and misinterpretation need to be addressed in subsequent warnings.

SOCIAL MEDIA USE BY THE PUBLIC DURING DISASTERS

With a growing fraction of the public using social media, there has been increasing interest in their use during disasters. One source of information on how social media are used, presented by Kristiana Almeida, comes from a 2011 American Red Cross study[3] in which approximately 2,000 people were interviewed. Nearly 50 percent of those interviewed reported that they visited one or more social media sites almost every day. For those in metropolitan areas, this percentage was higher. Although one in six subjects reported that they had used social media to find information about an emergency, television continues to be their primary source of information during these events. However, in cases where people did not have access to a television, Facebook became their primary information source. Furthermore, more than 80 percent of those interviewed indicated that they were willing to post information about a disaster on social media sites, including images or video clips.

More than a third of the regular social media users interviewed in the Red Cross study indicated that they would request help via social sites, and of those who would post requests for help, 80 percent expected a response within an hour. However, only 15 percent believed that emergency management agencies were actively following social media during emergencies. That is, although many believed that emergency officials should be following their feeds, few believed that they actually were.

Several key lessons could be taken from the 2011 Red Cross study, observed Almeida. First, it suggests that there are opportunities for emergency managers to use social media "crowdsourcing"[4] to supplement their situational awareness during emergencies. Second, because (according to the Red Cross study) public officials are not currently prepared to respond to requests for assistance via social media despite growing public expectations for such a capability, officials will need to continue to remind the public to use 911 to call for assistance.

BARRIERS TO INCORPORATING SOCIAL MEDIA INTO EMERGENCY MANAGEMENT

Although some emergency management organizations have begun using social media to interact with the public, several barriers and challenges remain to more widespread adoption, including limited understanding of social media, concerns about loss of control, and institutional limitations. Drawing on lessons he learned from discussions with fellow local emergency managers, Edward Hopkins, Maryland State Emergency Management Agency, noted the follow challenges:

---

[3] American Red Cross. Social Media in Disasters and Emergencies. 2011. Available at http://www.redcross.org/www-files/Documents/pdf/SocialMediainDisasters.pdf.

[4] *Crowdsourcing* refers to the use of information provided by multiple, and often many, individuals, who may have volunteered this information or been paid or given some other incentive to provide it. Chapter 3 explores the use of crowdsourcing to win a competition.

- *Limited understanding.* Despite growing appreciation of the role that social media can play in disasters, many emergency managers remain more comfortable with traditional media, and not all are aware of the potential advantages of social media as a tool for alerts and warnings. Some of this is surely a result of generational or cultural differences. Familiarity and comfort with social media for emergency management can be expected to grow as training opportunities are provided and newly hired employees bring with them a greater familiarity with social media. (The adoption of social media by practitioners is discussed further in Chapter 2.)
- *Loss of control.* When they use social media, emergency officials cannot control which information social media users share, which raises concerns that they might lose control of messaging or face civil liabilities if misinformation is shared. By contrast, in the traditional command-post style of information dissemination, long-standing relationships between the press and emergency managers provide some sense of control over what information is disseminated as well as well-understood opportunities to disseminate corrections as needed. It is the belief of some officials that with social media, misinformation may spread more rapidly and continue to spread even after a correction is issued.
- *Institutional limitations.* Especially in an era of shrinking budgets, it is hard to find the resources to evaluate and adopt new tools and technologies, or to invest in the training necessary for their use. Also difficult is securing new resources to cover the additional staff time necessary to monitor social media activity. As a result, although emergency management personnel may try to experiment with social media use, they may find that their other (day-to-day and emergency) responsibilities crowd out the possibility for this work. Also, such experimentation with social media often precedes the creation of guidelines or formal policies for their use, which can lead to unforeseen complications or questions being raised by managers, and implementation of measures that restrict the use of social media.

TECHNOLOGY DEVELOPMENT FOR THE USE OF SOCIAL MEDIA IN EMERGENCY MANAGEMENT

Emre Gunduzhan discussed technologies that are needed to disseminate public alerts and warnings using social media and to develop situational awareness during disasters using information gleaned from social media. Although some of these tools may be available now, they might not have been used in emergency management practices. Additional tools will also have to be developed that focus on the specific needs of disaster response, commented Gunduzhan, who also outlined the following set of technical challenges common across multiple alerting systems:

- *Standard message formats.* The Common Alerting Protocol (CAP) is a recently adopted protocol being used across multiple alerting platforms.[5] The advantage of CAP is that as an XML-based standard, it is machine-readable, and Web services and applications can receive and process these alerts. More recently, an Internet Engineering Task Force (IETF) working group, Authority-to-Citizen Alerts, has been developing a new protocol for transmitting alerts to the public over the Internet.
- *Capabilities for authorizing users.* Some sort of scheme, which might be a centralized, federated, or distributed identity management system, is needed to help ensure that only those properly authorized to do so can issue official messages.
- *Digital signatures that indicate who sent a message.* Public response is affected by who sends a message. The public needs to be able to verify who has sent a message and that the message is authentic. Nonrepudiation supports the assignment of responsibility for the issuance of an alert and protects from potential liability parties who relay an official message in good faith.
- *Geotargeting of alerts and warnings.* Controlling the granularity of alerting by geographically limiting the area targeted for receiving an alert might greatly protect the public from alert overload and better ensure that information gets to those who need to take a particular protective action (e.g., shelter in place versus evacuate). With respect to this capability, social media services diverge prominently from other alerting systems. In other existing communication channels, physical parameters, such as the area served by a set of cell towers or the metropolitan area served by a radio or television broadcast tower, constrain the distribution of alerts to a specific geographical area. By contrast, social media services embody limited and imperfect knowledge of the precise location of their users. Most often the information available is the address or town that may have been provided as part of a user's profile; a user's Internet Protocol address may also provide clues about his or her location. Although users accessing social media services via mobile devices may be able to provide accurate geotargeting information, users generally need to enable this feature, and many choose not to do so owing to privacy concerns.

There are two possible methods for using social media to deliver alerts and warnings. One is for the entity issuing the messages to simply be registered as a user of the service, as would any other information provider. The other is for the entity to establish a special relationship with

---

[5] OASIS. Common Alerting Protocol, v. 1.1. OASIS Standard CAP-V1.1, October 2005. Editors: Elysa Jones and Art Botterell. Available at http://www.oasis-open.org/apps/org/workgroup/emergency/download.php/14205/emergency- CAPv1.1-Committee%20Specification.pdf.

the social media service. The characteristics of each method are outlined in Box 1.1.

Today, emergency managers are making use of only the first alternative, noted Gunduzhan. For example, a local jurisdiction may set up a Twitter account and simply post warnings as an update that would be seen by those who follow that account or find the post as the result of a search. Similar possibilities exist with other social media services like Facebook and Google+, although who sees what messages and under what conditions is more complicated owing to the service design. Such arrangements have the advantage of being possible to implement today without reaching agreement on a new interface to deliver alerts and warnings or new arrangements for their display to users at risk from an event. On the other hand, they require that emergency managers find a set of clients that allow them to post messages to each service, and no special provisions are available to ensure that affected populations see the right set of alerts and warnings. (There is also a question of how to establish the authenticity of alerts and warnings; see above.)

Alternatively, emergency managers might establish partnerships with social media services for disseminating alerts and warnings, either on behalf of individual agencies or collectively at the state or federal level.

BOX 1.1
Characteristics of Alternative Methods for Distribution of Social Media Alerts

Distribution by a Registered User of Social Media Services
- Alerts appear as ordinary messages.
- Alerts are difficult to scale and maintain due to many different application programming interfaces (APIs) that are subject to change at any time.
- Citizens have to register for the service (opt in).
- Alert authentication and authorization are managed externally to the social media site.
- Non-repudiation is not a major issue.

Distribution Via Collaboration with Social Media Sites
- Alerts can be given priority and special treatment.
- A single standard interface definition can be used by all collaborating social media sites.
- Citizens will automatically be included in the service but may be allowed to opt out.
- The social media site implements additional alert authentication and authorization.
- Non-repudiation and liability protection need to be addressed.

For example, social media providers could agree with emergency managers to take steps in concert to make alerts more obvious to users. Such an approach is analogous to the Commercial Mobile Alert Service, in which cellular carriers have been working with the federal government to establish a standard message format, alert tone, and national gateway for delivering messages.

TECHNOLOGIES FOR DEVELOPING SITUATIONAL AWARENESS FROM SOCIAL MEDIA

Several workshop participants observed that unlike other means of delivering alerts and warnings, social media technologies offer the distinct advantage that they are two-way, are interactive, and provide an opportunity to see how the public is responding to the message. That same interactivity poses new challenges, however, such as managing the sheer volume of messages that can be sent during a major event. For example, following the announcement of Osama Bin Laden's death, messages ("tweets") were being posted to Twitter at a rate of about 3,000 per second.

Given the flood of information available from social media, some workshop participants observed that automated filtering tools are likely to be key to developing situational awareness. For example, automation can help categorize sources and separate relevant and irrelevant information. Progress toward such automated tools will rely on advances in natural-language processing. The location of a source of information is also helpful in separating out relevant information and can sometimes be determined from a user's profile or from metadata associated with the message, or inferred from the content of the message itself. Several workshop participants noted that even with greater automation, human judgment will be needed to interpret and act on the selected messages.[6]

Visualization tools that can help emergency managers make sense of social media information are already available to some extent, observed workshop participants. Ushahidi, an open source tool for information collection, visualization, and interactive mapping,[7] for example, has been used successfully during a number of disasters.[8] Looking ahead, one

---

[6] In an effort to combine technological tools and human judgment to monitor social media, the American Red Cross opened its Digital Operations Center in March 2012.

[7] See http://www.ushahidi.com/.

[8] The value of Ushahidi during the 2010 earthquake in Haiti is explored in Nathan Morrow, N. Mock, A. Papendieck, and N. Kocmich, *Independent Evaluation of the Ushahidi Haiti Project*, Development Information Systems International, 2011, available at http://www.alnap.org/pool/files/1282.pdf. The report explains, "Perhaps the most common use of information aggregated by UHP was for situational awareness. The Department of State

challenge is how to integrate information from multiple social media services as well as from other information sources. Gunduzhan noted that addressing this challenge will require attention to a mix of standard interfaces and formats as well as translators.

The use of tools to gather and analyze information derived from social media services raises a set of privacy issues associated with the collection, processing, retention, and distribution of such information. These issues were explored in another workshop panel, whose observations are presented in Chapter 5.

OBSERVATIONS OF WORKSHOP PARTICIPANTS

During the discussion, panelists and other workshop participants offered a number of observations on the use of social media for alerts and warnings, including the following:

- Although the public response to alerts and warnings has been studied for some time, as have more general questions about the information needs of the public during disasters, there has been comparatively little research on the use of newer technologies such as mobile devices (e.g., cell and smart phones) or social media during disasters. Research would help shed light on such key questions as how the new technologies could help shorten the milling time between receipt of an alert or warning and the taking of protective action.
- Questions still remain regarding the extent to which social media represent a truly new source of information for improving the situational awareness of emergency managers or whether much or most of the information simply repeats information already available from other, traditional sources.
- The particular characteristics of social media platforms yield different "affordances" of use during disaster situations. For example, social networks on Facebook are usually user-defined: messages to one's network might help target localized attention and allow extended discussion. Facebook newsgroups, however, are public and can draw collected attention. Twitter supports rapid communications that are most often public. The communications are short and can easily be "retweeted" or

---

analysts for the USG interagency task force used Ushahidi in at least one case to help triangulate conclusions about the situation on the ground, and US military organizations used Ushahidi data feeds along with other sources in a similar manner to inform their early situational assessments. There is also some evidence of the information being used for specific operational and tactical actions targeting specific communities (and to a much lesser extent, individuals). US marines used the information to identify 'centers of gravity' for deployment of field teams to [affected] areas."

propagated publicly. Those that are propagated have a chance of receiving attention; those that are not die out quickly.

- It is important not to think of these platforms as better or worse; rather, it is critical to understand that they are different places along sometimes circuitous paths to seeking and finding information.

2

Current Uses of Social Media in Emergencies

A variety of emergency management organizations and other groups that communicate with the public during disasters have been using social media to disseminate information and observe the public response to events. In this workshop session, four speakers shared firsthand experiences with the use of social media: Brian Humphrey, Los Angeles Fire Department; Brad Panovich, WCNC-TV in Charlotte, North Carolina; Paul Earle, United States Geological Survey National Earthquake Information Center; and Keri Lubell, Centers for Disease Control and Prevention.

USE OF SOCIAL MEDIA BY THE LOS ANGELES FIRE DEPARTMENT

Brian Humphrey described the wide array of social media tools the Los Angeles Fire Department (LAFD) uses to both disseminate and monitor information before, during, and after fire emergencies, including two Twitter accounts: @LAFD,[1] used solely for alerts, and @LAFDtalk,[2] used for conversations with the public. The latter account is used to help build trust between the public and the fire department. For example, when followers share personal news, such as birthdays or anniversaries, via Twitter, @LAFDtalk may acknowledge these occasions to engage with fol-

---

[1] See https://twitter.com/lafd.
[2] See https://twitter.com/lafdtalk.

lowers socially. Status updates often include names or initials to reinforce that there are real people behind the account.

The value of the LAFD's engagement of the public through social media was highlighted during an incident in which an explosion was reported at the Los Angeles International Airport. The LAFD could see who had recently indicated via Foursquare[3] that they were at the terminal where the explosion was reported, and asked these individuals (via Twitter) to contact its public affairs office by telephone. By asking these individuals what they had observed, the public affairs office could provide first responders with information about the event even before they were on the scene; in this case, the public affairs office was able to tell the responders that the explosion was in fact the result of a lithium battery overheating, a relatively minor incident.

Nevertheless, some potentially useful tools for using social media have proven too expensive to acquire, and some uses of certain tools are precluded by the terms of their end user license agreements, commented Humphrey, but he noted that the LAFD has identified a set of free or low-cost tools that are useful for disseminating and monitoring information. Many of the LAFD's social media accounts are fed by email using the services Ping.FM[4] and HelloTXT.com,[5] which makes it possible to quickly provide or request information through the approximately 80 social media accounts managed by the LAFD. These accounts, including @LAFD and @LAFDtalk, can also be used individually to interact with the public.

Different social media tools are useful during different stages of a disaster. For example, according to Humphrey, Blog Talk Radio[6] is a tool that is particularly helpful during the recovery stage. Blog Talk Radio can be used to create Internet radio stations, which can take calls from users. Other related tools are also useful. Standards such as RSS (Really Simple Syndication) and various XML (Extensible Markup Language) schema make it easier to represent, distribute, and analyze information.

One task for which better tools would be helpful, Humphrey observed, is in extraction of information from photographs distributed over social media. Currently, the LAFD relies on manual searches for potentially useful pictures based on the social media tags they are associated with. Another possible source of information is the metadata included in image files, such as the time or location. It would also be helpful to have tools

---

[3] Foursquare is a location-based, social media application for mobile devices. Users "check in" to locations found near their current location, which is detected using the GPS hardware in the mobile device. See https://foursquare.com.

[4] See https://ping.fm.

[5] See http://hellotxt.com/.

[6] See http://www.blogtalkradio.com.

that could automatically identify pictures that might enhance the LAFD's situational awareness about events, such as pictures that show fire or smoke.

Humphrey concluded his remarks by identifying several important lessons from the LAFD's experience with social media:

- A partnership with information technology (IT) staff in the emergency management organization is important in part so that IT staff understand how emergency professionals are using the computers and network.
- Appropriate management of and collaboration with traditional and new media are critical components of each phase of a given disaster. There is often a gap between response (usually an acute situation) and recovery (a more ongoing process). Social media can smooth the gaps in the cycle by providing two-way communication between the affected population and both first responders and emergency personnel.
- Messages disseminated using social media need to be clear, concise, and, most important, actionable.
- The more one is willing to empower people and engage them, the more information the public is willing to provide.
- Understanding how people communicate with social media is important. People often do not simply state "help," "fire," or "explosion" but instead use such exclaimers as "OMG!" (oh my god!) or other slang. When one has three or more people within a 20-mile radius saying "OMG," this can be a signal to look more closely at what might be happening in the area.
- Different phrasing suggests quite different meaning, as the following examples regarding a fictional shooting illustrate:
    — I *heard there was* a shooting at 5th and Elm St.
    — I *heard there is* a shooting at 5th and Elm St.
    — I *heard about* a shooting at 5th and Elm St.
    — I *heard* shots fired at 5th and Elm St.
    — I *saw* a shooting at 5th and Elm St.
    — I just *saw a person* get shot at 5th and Elm St.

USE OF SOCIAL MEDIA BY WCNC, CHARLOTTE, TO PROVIDE WEATHER INFORMATION

Brad Panovich began by describing the multi-tiered warning process he uses to inform the public when potentially severe weather is forecast. The first step is a blog post several days in advance to increase public awareness. As severe weather approaches, Panovich begins issuing alerts or warnings via Twitter and Facebook. Many of the alerts and warnings originate from iNWS, a service of the National Weather Service, which sends information via text and email to emergency managers, local county officials, and the media.[7]

Panovich, using his desktop computer, laptop, or other mobile device, needs to be able to quickly post alerts and warnings from iNWS to his social media accounts. In Panovich's workflow, these messages are sent to an email account that has a rule set to automatically forward iNWS messages to TwitterMail, which in turn posts the alerts to Twitter. Panovich noted that a more complicated setup is needed to automatically post messages to Facebook because only the subject of an email is included in a status update when a message is emailed to Facebook. Instead, tools like Ping.FM and Tumblr[8] can be used to automatically post information to Facebook. A text alert from iNWS can be pasted into Ping.FM, which is configured to send the message to any of the six Facebook pages that Panovich manages. Tumblr blogs allow for posts via email, and then these posts can be sent directly to an RSS feed, a Web-based tool that can display various updated Internet sites in a standard format, or automatically posted to Facebook or Twitter.

WCNC uses YouTube and uStream to transmit video over the Internet. One use is to supplement the broadcast forecasts with additional background on the science behind the weather, which is appreciated by audiences and helps strengthen Panovich's credibility with them. Streaming also allows the public to view forecasts and other information on computers and mobile devices[9] and can be used as a supplemental source of information to point people to an alert displayed over normal television programming, allowing viewers who seek additional information to quickly find it using a cell phone or other mobile device. In general, Panovich believes, the public finds this approach less disruptive than interrupting programming to provide information to supplement an alert, especially when many of those receiving the broadcast may not be in the affected geographical area.

Panovich's viewers and readers access his information using both traditional and social media channels. Understanding an audience's needs is important, Panovich observed. He uses a tool called SocialBro to analyze the activities of his followers. For example, information on when users are most active helps inform decisions about when to create new content. The

---

[7] See http://inws.wrh.noaa.gov/.
[8] Tumblr is a social networking site that allows for the sharing of microblogs, reposting of multimedia, or writing of short blog items. See https://www.tumblr.com/.
[9] To provide mobile device streaming, it is important to use the correct codex for mobile devices.

tool also can be used to understand other aspects of users, such as what words are used to describe significant weather events.

Panovich identified several important lessons from his experience using social media as a broadcast weather forecaster:

- It is important to build a social network before an emergency occurs.
- Although automatically transmitted alerts are useful, the best results come through a dialog with the public.
- Because people use social media around the clock, a forecaster must be quite active to fully engage the public.
- Social media and other new tools for disseminating information are important, but not everyone uses social media, so broadcasting and other traditional media also play an important role in informing the public about weather events.

USING SOCIAL MEDIA FOR EARTHQUAKE DETECTION AND ALERTING

The National Earthquake Information Center (NEIC) monitors earthquakes worldwide. Additionally, the center provides backup for the network of regional seismic centers. The NEIC's customers include relief organizations, government agencies, and research and financial institutes, as well as the media and general public. The main products include real-time monitoring (which is used mostly by researchers), estimates of the magnitude of ground shaking that are based on seismic monitor readings and damage reports, and rapid estimates of fatalities and economic damage on a green-yellow-orange-red scale. The rapid estimates are emailed to approximately 200,000 subscribers and are posted on the NEIC Web site. The NEIC also uses Twitter to disseminate alerts and detect earthquakes.

Paul Earle described an event detector developed by the NEIC that examines tweets[10] to detect when an earthquake has occurred. It watches for a sharp increase in the rate at which keywords associated with earthquakes are used. This tool detects between 1 and 4 earthquakes a day worldwide (seismic instruments detect approximately 50 a day). Although not as accurate as seismographs in determining either magnitude or location, it extends coverage in areas with little instrumentation and also provides a backup should instruments fail or be off-line. Where detectors are sparse, earthquake detection based on instrument readings can take up to 5 minutes, whereas detection using Twitter can take less than 2.

---

[10] Status updates posted on Twitter are often referred to as "tweets."

Earle described several limitations to Twitter-based detection. First, fewer than 5 percent of all tweets used are accurately geocoded. Second, the keyword detection algorithms have approximately a 10 percent false rate. Detection thresholds can be positively adjusted, but there will always be tradeoffs between false positives and false negatives.

To distribute earthquake alerts using Twitter, the NEIC manages a verified[11] Twitter account, @USGSted (US. Geological Survey Tweet Earthquake Dispatch). Although the Twitter feed currently reaches fewer people than does NEIC's email distribution list, the alerts are distributed almost instantly (whereas it can take up to 20 minutes to send emails to the approximately 200,000 email subscribers), and the Twitter feed is not as hard to manage as an email list.

The Twitter alerts contain the region name and geocoded information if available, the data, the time, and the rate at which Twitter messages used to detect the earthquake were sent. The Twitter messages do not include the quake magnitude because this information can change rapidly as measurements are verified, and retweeted messages may continue to circulate long after their content is no longer current. Instead, the messages contain a link to a Web page that provides current information on the earthquake. An example of a Twitter alert and of the Web page it links to is shown in Figure 2.1.

USING SOCIAL MEDIA TO ASSESS COMMUNICATION NEEDS AND DISSEMINATE INFORMATION DURING A HEALTH EMERGENCY

The mission of the Centers for Disease Control and Prevention (CDC) includes the detection of emerging health threats, informing the public about these threats, and informing public officials about potential responses and the associated risks and benefits. Sometimes this involves circumstances where knowledge about a potential threat is incomplete and where there is considerable uncertainty about its implications.

The CDC uses social media both to help detect emerging threats and to disseminate information to the public about how to respond, explained Keri Lubell. Social media can provide clues not only about emerging events but also about how people are responding to those events and to the information that the CDC and others are providing. As a result, there

---

[11] Verified Twitter accounts are marked with a blue verified badge. Verified users have authenticated identities, and verified accounts are generally not available to the general public but instead to those who are considered high-quality sources. A list of frequently asked questions on verified accounts is available from Twitter at https://support.twitter.com/articles/119135#.

FIGURE 2.1 An example of a tweet and the associated Web page for a detected earthquake.

is interest in automated approaches that could take greater advantage of all the potential information available from social media.

In terms of the use of social media to disseminate information, Lubell said that the CDC had been an early adopter but that the agency's use of social media and associated policies are still evolving. The CDC's Facebook page for emergencies (https://www.facebook.com/cdcemergency) has approximately 14,000 "likes,"[12] and the main CDC Facebook page (https://www.facebook.com/CDC) has about 210,000 likes.

The Twitter handle @CDCemergency was established in January 2009 and was first used during the 2009 *Salmonella typhimurium* outbreak that was associated with peanut butter. That account has approximately 1,350,000 followers, and the general CDC account, @CDCgov, has approximately 97,000 followers. (Up until NASA launched the last shuttle, the CDC had the highest number of followers in a single government Twitter account, Lubell noted.) If the agency can determine good ways to mobilize them, 1.3 million followers could prove a useful resource.

A recent use of social media in conjunction with the CDC's worldwide polio eradication effort illustrated both the public's interest in and potential pitfalls of using social media, observed Lubell. As part of this push, the CDC had activated its emergency operations center for an 18- to 24-month period. As a part of the campaign's communication strategy, the CDC along with its global partners used Twitter to engage the public and volunteers across the world. (Such prearranged public discussions are frequently used by the CDC to engage the public in important health issues.) During the discussion, several items were retweeted by @CDCemergency, with the goal of reaching the wider audience connected to @CDCemergency. However, within a few minutes, followers began complaining that because these messages about polio were not about an emergency, they were not an appropriate use of that feed (Figure 2.2).

In response, the @CDCemergency team provided a brief response and stopped using that channel. Lubell observed that the incident demonstrated that people who followed @CDCemergency were engaged and that it was important to adapt in the face of public response. It also pointed to the need for organizations like the CDC to develop strategies for how best to use the various social media channels at their disposal.

---

[12] A Facebook "like" is similar to "friending" or "following." By liking a Facebook page, a user may receive updates on the liked organization on his or her news feed. The information frequently changes. See https://www.facebook.com/cdcemergency or https://www.facebook.com/CDC for an up-to-date count of "likes."

> CDCemergency CDCGlobal WHO Rotary GatesPolio unfoundation ShotatLife EndPolioNow Thought this was emergency?
>
> Arghhhh! Why would CDCemergency RT a routine poliochat. Nothing emergent about it!
>
> CDCemergency please stop with the NON-emergency retweets!
>
> CDCemergency Isn't by definition "EMERGENCY" meant to keep the communication to a minimum unless necessary. Stop the non-ER tweets and RTs!
>
> CDCemergency These RTs are not regarding an emergency. Please stop. I only follow you in case of a smallpox outbreak.
>
> CDCemergency What's with all the tweets. I thought you were only supposed to tweet in an emergency. I'm close to unfollowing you
>
> CDCemergency too many non emergency tweets = unfollow. Thought you should know
>
> CDCemergency do you have to spam my twitter account with the same message several times today.
>
> The CDCemergency account appears to indicate there is either a polio emergency or it was hijacked by someone who really likes polio
>
> Nice to see CDCemergency abusing their "emergency" twitter channel for non-emergency info. Great way to gain credibility! sarcasm

FIGURE 2.2 Examples of the public's response to the use of @CDCemergency to send messages about the Centers for Disease Control and Prevention's global polio eradication campaign.

OBSERVATIONS OF WORKSHOP PARTICIPANTS

Observations on current uses of social media offered by workshop panelists and participants in the discussion that followed the panel session included the following:

- Emergency response agencies, notably in the areas of fire and public health response, have been successful in building considerable trust with the public. There are potential risks involved in using social media, and government agencies tend to be very risk-averse. However, social media can significantly enhance public response capabilities and can also be used to enhance public trust.
- Social media offer a partnership through which officials can provide the best information they have available at the moment and their followers can help disseminate the information.

- The public response to events, including both individual messages and trends such as increased social media traffic in a location or about a particular topic, can help responders understand an event as it unfolds.
- Local officials, who already have credibility advantages because they are perceived as being "in the same boat" as the public they serve, have had success in fostering public trust in advance of crisis events and in using social media to communicate effectively with the public during such events.
- As communication becomes increasingly mobile and Internet media services are substituted for broadcasting services, social media applications can provide an increasingly important way for emergency managers to reach the public.
- Although a number of existing tools have been successfully adapted, there is also a need for tools for both information dissemination and monitoring that are better matched to the needs of emergency managers.
- The use of social media for disaster response requires significant advance planning. This includes experimenting with various workflows and technologies to assist in the rapid dissemination of information.

3

Dynamics of Social Media

How social connections form, how information is disseminated within social networks, and why people volunteer their time and knowledge to solve problems are all questions that have been examined by researchers in sociology and computing and experienced by those using social media to coordinate disaster response. Duncan Watts, at Yahoo! Research at the time of the workshop and since at Microsoft Research, discussed some of what researchers have learned about how people use social media and the implications for use of social media in disasters, drawing on research about Twitter users. Manuel Cebrian, University of California, San Diego, examined approaches for incentivizing participation in time-critical situations, drawing on lessons from two recent challenges sponsored by DARPA. Melissa Elliott, Standby Task Force, discussed the dynamics of social media during a crisis, drawing on experience with volunteer efforts to use social media for disaster management.

STUDYING TWITTER USE TO UNDERSTAND HOW PEOPLE COMMUNICATE

In the United States, hundreds of millions of people interact with media sources and each other via social media, making the number of nodes and connections in an entire social media network incredibly large. The enormous diversity in the subjects being discussed via social media and the range of effects are even harder to study. Social media, and the data they yield about people's interactions, have emerged as a valuable new lens through which to explore the full range of communication among individuals. Duncan Watts discussed work at the intersection of social science and computer science performed at Yahoo! that used a subset of Twitter user information and updates.[1]

Media research has tended to focus on two types of communication—individual organizations broadcasting to large, undifferentiated audiences, and individuals communicating with each other—but generally has not looked at anything that happens between these two extremes. Although most people think of Twitter as a social network, it can also be viewed as a full-spectrum media ecosystem.[2] Twitter communications cover the spectrum between the two types of communication traditionally examined by media research; individuals as well as traditional mass media outlets are able to broadcast information. New forms of interaction have emerged, such as mass personal communication, in which "elite" individuals—celebrities, politicians, journalists, or recognized experts—not only broadcast information to large audiences but also engage in public conversations that are widely followed.

One of the biggest challenges to using social media, Watts noted, is the large number of accounts and the volume of data they generate. It is difficult to categorize the more than 200 million Twitter accounts as those associated, for example, with individuals or organizations. In 2009, Twitter introduced a new feature called lists, which provided users with a mechanism for filtering incoming feeds and other users, providing researchers with data (which is public by default) on how users classify each other.

Watts explained that the Yahoo! study drew on a collection of data originally used by Haewoon Kwak in his study of Twitter. Collected in 2009, the data included 42 million users and 1.5 billion individual connections.[3] (However, the focus of the work was on 260 million tweets that included a bit.ly URL, a URL-shortening service.)

One important finding from this work was that a small number of "elite" users were followed by half of all Twitter users. Yahoo! used the list feature to help separate out four categories of elite users: celebrities,

---

[1] Shaomei Wu, Jake Hoffman, Winter Mason, and Duncan Watts. Who Says What to Whom on Twitter. 20th Annual World Wide Web Conference, Association for Computing Machinery, Hyderabad, India, 2011. Available at http://research.yahoo.com/pub/3386.

[2] Haewoon Kwak, Changhyun Lee, Hosung Park, and Sue Moon. What is Twitter, a social network or a news media? Available at http://product.ubion.co.kr/upload20120220142222731/ccres00056/db/_2250_1/embedded/2010-www-twitter.pdf.

[3] Haewoon Kwak originally made the data public at http://an.kaist.ac.kr/traces/WWW2010.html. However, a change in Twitter's terms of service resulted in the researchers being unable to share their original data set.

media outlets, organizations and corporations, and bloggers.[4] Researchers put Twitter users into one of the elite categories based on how frequently they were categorized as such by individual users. For example, a Twitter handle labeled as that of a celebrity by 20,000 other users most likely did in fact belong to a celebrity. What was learned was that 50 percent of all attention was being paid to just 20,000 elite users. This is not to say that those elite users were producing half of all tweets, but rather that half of all the tweets that were read were updates provided by these elite users.

Focusing on the above four categories of elite users suggests an interesting corollary to what is known as the homophily principle in sociology. People who are connected are more similar than people who are not connected. Celebrities pay attention to other celebrities, the media follow the media, and so on. Only users in the organization category pay more attention to users in other categories than to users in their own category. Several corporations, non-governmental organizations, and government organizations are using Twitter to listen as much as to talk. With retweets, the pattern is ever more striking. Celebrities rarely retweet messages from anyone. Bloggers, on the other hand, do a tremendous amount of retweeting, which is consistent with the stereotype of a blogger as a synthesizer and distributer of information.

Another important finding cited by Watts was that a relatively large fraction of the population received information indirectly. To examine the flow of information, researchers studied the propagation of URLs that originated from the media category, which consisted of about 5,000 accounts. Approximately half of this information reaches users indirectly. A lot of information did not come directly from the media source but instead indirectly through other accounts, which were labeled opinion leaders by researchers. The number of opinion leaders was incredibly large. They were consuming much more content than normal users but also tweeting more and had a higher number of followers. These results suggest that many social media users will receive alerts from a non-authoritative source.

A related issue is what sources of information have the most influence. Watts and his team used retweets as a measure of influence; they assumed that an individual who was frequently retweeted was probably more influential. By tracking tweets and retweets, they were able to develop influence trees that described how a tweet was cascaded through the Twitter ecosystem. It turned out that most URLs included in Twitter messages were not retweeted by anyone and that the average number

---

[4] Shaomei Wu, Jake Hoffman, Winter Mason, and Duncan Watts. Who Says What to Whom on Twitter. 20th Annual World Wide Web Conference, Association for Computing Machinery, Hyderabad, India, 2011. Available at http://research.yahoo.com/pub/3386.

of retweets was around 1.2. This is a relatively small number, especially considering the vast literature on diffusion theory that predicts very large cascades. However, very large cascades were also observed. For example, the data examined by Watts et al. included at least one cascade of about 10,000 retweets. Another important factor is how many generations of retweets occur. Several cascades extended several generations, with information moving farther away from the source, but almost 90 percent of retweets went only one step away from their origin.

Two things about influence can be inferred from the Twitter data, Watts observed. First, someone with influence probably will continue to be influential, and second, the number of followers someone has increases his or her influence. No other factors were found to affect influence. For example, the content of the tweet was in general not predictive of how influential it was (or how many retweets it received)—although this finding may not hold true in particular contexts such as emergency events in which the value seen in sharing critical information may be higher.

PROBLEM SOLVING WITH SOCIAL MEDIA

In addition to seeing social media as a source of information in disasters, it is also natural to consider how social media can be used to engage people in solving problems. The essence of the question is how one can use social media and the right set of incentives to engage people to solve a set of tasks for which humans are well suited. Recent work by Manuel Cebrian in the context of two online challenges mounted by the Defense Advanced Research Projects Agency (DARPA)—the DARPA Network Challenge and the DARPA Shredder Challenge—has provided some empirical knowledge and further insights. The first contest, the 2009 DARPA Network Challenge, offered a $40,000 prize to the first team to find red weather balloons placed in 10 undisclosed locations in the continental United States.

A key design issue facing the teams, according to Cebrian, was how to recruit participants. The winning team, from the Massachusetts Institute of Technology, chose a variant of the query incentive network model, first developed by Jon Kleinberg and Prabhakar Raghavan in 2005.[5] The individual who actually found a balloon would receive the largest award, but those who recruited that individual would also be awarded an amount equal to half of what the connected award winner received. For example, if Dave found a balloon, he would receive $2,000; Carol, who recruited Dave, would receive $1,000; Bob, who recruited Carol would receive $500;

---

[5] J. Kleinberg and P. Raghavan. Query incentive networks. *Proceedings of the 46th IEEE Symposium on Foundations of Computer Science,* 2005, pp. 132-141.

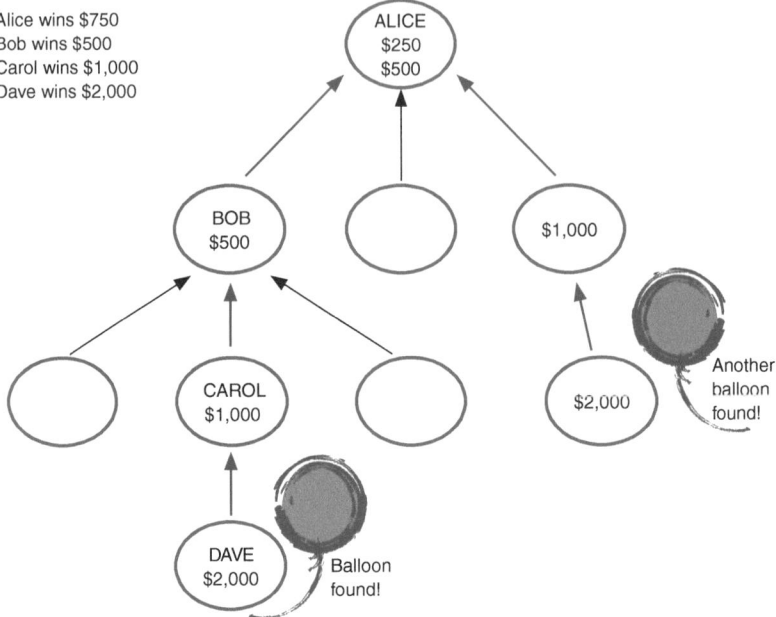

FIGURE 3.1 Recruitment and award distribution. SOURCE: Manuel Cebrian. Social mobilization under "the fog of war." Presented at Workshop on Alerts and Warnings Using Social Media, February 28-29, 2012.

and Alice, who recruited Bob, would receive $250. Figure 3.1 illustrates how awards would be distributed in this scheme.

In roughly 8 hours, the MIT team recruited approximately 4,500 participants. Recruiting started in major cities and then spread into the suburbs, which the team believes ultimately played a role in finding some of the very-difficult-to-locate balloons. With such a large number of recruits, the team expected to simply locate the balloons and win. Recruits would submit possible locations, the team would examine the density of submissions, and the prominent 10 locations would coincide with balloon placement. However, in the first day of the competition, of the 400 submissions received, 85 percent were incorrect, and it became clear that people were attempting to sabotage the team's effort. Initially the false locations were simply random, but later it became clear that some of the spoofing was being coordinated to provide multiple reports of the same location.

The MIT team developed several techniques to filter out the false reports. One was to question multiple identical locations; correct submissions were more likely to contain very close but not exactly identical locations. Another technique was to discount reports from someone located far away from the reported site. The increased rate of incorrect reports also led the team to require a photograph for verification. Spoofers then began fabricating pictures, which led to a requirement that the photographs show the DARPA representative stationed at each balloon as well. That in turn led to someone dressing up as a DARPA official in order to submit misinformation.

Ultimately, the MIT team was able to win because it had a small number of highly motivated participants who physically visited sites to visually confirm the presence of a red balloon. Several questions arose from the challenges presented by intentional misinformation, and in an attempt to test a set of related hypotheses, the same research team attempted to design the incentive network for the next DARPA challenge in a similar way.

The DAPRA Shredder Challenge asked individuals or teams to reconstruct a progressively harder series of puzzles consisting of documents that had been shredded into fragments. The first puzzle had 200 pieces and the fifth had 6,000 pieces. By comparison, the best computer algorithms are able to solve a 400-piece puzzle, which meant that computational approaches alone could not be used to win the challenge. The first team to reconstruct the puzzle would win $50,000.

In an attempt to mimic the red balloon challenge incentive, MIT researchers decided that each of the pieces for all five puzzles in the challenge would correspond to a $1.00 award for correctly placing the piece. Once again, the network of the individual correctly placing a piece would also be rewarded. If an individual received $50.00 for assembling 50 pieces, the person who recruited the individual would receive $25.00. Over 2 weeks, 3,500 people joined the team. Within those 2 weeks numerous teams had already solved the first two puzzles. After 4 days the MIT team rose to third place in the competition by solving the first three puzzles.

In the first three puzzles, steady progress was made toward a solution, but starting with puzzle four there were slowdowns in progress owing to intentional sabotage. Notably, those responsible for sabotaging the effort were also the same individuals who earlier had contributed solutions to hard pieces of the problem, making it difficult to filter out the bad actors.

One saboteur contacted Cebrian, confessed to damaging the puzzle, and provided a summary of the techniques used to create the damage. The individual first enlisted the help of members of an online bulletin board to disconnect puzzles. In response, Cebrian locked correctly placed pieces in place and banned certain Internet Protocol (IP) addresses. The malicious individual then began using a virtual private network and open wireless networks to appear from a new IP address and was able to again damage the puzzles. Other users quickly began noticing that the malicious individual was simply stacking pieces; the individual then simply began moving pieces off the virtual workspace. As for motivation, his claimed rationale was that the shredder puzzle was intended to be a computing/programming challenge and that crowdsourcing was "cheating."

Cebrian's team approached the problem of saboteurs in several ways, most of which turned out to be mistakes. First, it attempted to police what was happening and to find the saboteurs. Although the team was able to determine that a group in San Francisco and a group in Amsterdam were coordinating the attacks, this information did not contribute to solving the puzzle. A second mistaken approach was to restrict participation in an attempt to block malicious people—initially, the virtual table had been available to everyone to move pieces at will. An additional step was to limit new users to moving one piece every 3 minutes. Ultimately, when the team reached the fifth puzzle, only the top 20 performers were allowed to participate. However, the fifth puzzle was simply too complicated for even the best performers to solve, and the team received no points for puzzle four and puzzle five and ultimately placed sixth place in the competition.

Both the red balloon and the shredder challenges demonstrated that one can recruit a large crowd to solve a very hard problem, and the analytical tools to understand how that can happen do exist, observed Cebrian. However, when competitive forces arise, it can be very difficult to determine the origin of those forces and what others' goals are. During time-critical situations, there is little leeway to contemplate the best way to thwart the impact of malicious individuals, and it becomes easy to feel paranoid and limit participation. The combinatorial nature of the shredder challenge as compared with the red balloon challenge (solve versus search) made social mobilization much more problematic. Searching for a single balloon has no impact on the search for a second balloon; however, in the puzzle challenge, each step built on a previous step, and the usefulness of crowdsourcing appears to have degraded.

STANDBY TASK FORCE: VOLUNTEER NETWORKS DURING DISASTERS

Melissa Elliott is a core team member of the Standby Task Force and is also a member of both Crisis Mappers and Crisis Commons. All three of these volunteer organizations work to coordinate volunteers who develop information-sharing tools and provide information to relief organizations during a disaster.

One of the most difficult aspects of using social media during the 2010 Haitian earthquake, according to Elliot, was the lack of processes and systems to coordinate information. Essentially, the disaster management teams, both non-government organizations and officials, were receiving multiple messages from multiple people. The repeating of old distress messages became a significant problem. Verification of shared information presented a growing challenge.

In response, the Standby Task Force was launched at the 2010 International Conference of Crisis Mappers in Boston. The purpose of the task force was twofold: to provide predictable crisis-mapping support to humanitarian organizations and to create a model for volunteer engagement according to a set of processes so as to maximize efficiencies and minimize redundancies. Today, the Standby Task Force comprises almost 800 volunteers representing 60 different countries. Not every volunteer will participate in a deployment (although the organization is looking for ways to increase volunteers' involvement). Many have joined the group as observers to understand how the task force works, ideally before participating more actively.

In January 2012, the Standby Task Force used a survey to help determine what motivates its volunteers and found that 45 percent participate because they are generally concerned about the people they help. Another 35 percent volunteer for the experience and to gain skills in technology use and hands-on crisis response. Several became involved to gain visibility outside their immediate geographical area.

The task force is divided into several teams, each of which focuses on a particular task, including analysis, geolocation, humanitarian aid, media monitoring, reports, satellite imagery, mobile text messaging, translation, verification, and technology support. The teams find, map, verify, curate, and analyze different forms of social media to improve the situational awareness of responding organizations. Volunteers join particular teams but often cross-train in multiple teams. Training and coordination are done online via a closed platform, Ning,[6] and using Skype videoconferencing. During deployments, a wide variety of tools are used; however, 80 to 85 percent of deployments are done using Ushahidi. Google Maps and Apps and Open Street Maps are also used.

Over the last 2 years, the organization has had 18 deployments. Deployments are incredibly time intensive for volunteers, especially the coordinators. Volunteers' work schedules are created and managed very closely to ensure that volunteers do not work continuously and that they take appropriate breaks from the work. The schedules also provide a way for participants to take ownership of their deployments.

The beginning of each deployment is fairly chaotic, Elliott observed. Volunteers are eager to get started; however, coordinators need to ensure

---

[6]See http://launch.ning.com/.

that they are all moving forward on the same path. Before beginning, teams establish workflows, schedules, and team-specific technology platforms to manage the process (mostly via Skype).

Initially the chat window fills up with brainstorming. Workflows, the process by which the task force filters information through the team, will not change. However, during a deployment, the way in which data is gathered may shift as volunteers leverage their contacts, share data sets, or use a specialized skill or tool for data visualization that the task force might not have thought to use. Maintaining flexibility continually reinforces a sense of ownership. However, coordinators are also continually reinforcing the workflow process by which the information is filtered.

Training is an ongoing process. Volunteers may have joined the task force between formal training sessions. In this case they receive extra support and, if time permits, may receive a one-on-one speed training session on whatever tool is being used during a deployment. Within hours of a deployment, members of the group begin stepping into leadership roles by providing direction and mentorship to new volunteers just joining. This approach allows coordinators to take a step back because they no longer need to be online constantly answering questions and reinforcing the process.

The deployment teams are composed of digital volunteers who all want to support one another during the challenge of deployments. As this support builds, more trust develops within the group. A self-correcting process begins as well. For example, if the media-monitoring team (which is the largest team and does much of the data mining of social networks) creates a report in Ushahidi and the geolocation data is incorrect, this error can quickly be recognized and corrected by other volunteers. In addition, if an individual submits several erroneous reports, he or she is able to receive additional training immediately. Sharing the burden of ensuring correct information creates another avenue for giving ownership to the crowd for that information and continues to reinforce a need to mentor and help others who may submit data incorrectly.

The large amount of data that arrives via Twitter, for instance, requires some initial sorting and then verification. Approximately 80 percent of the information the organization receives from Twitter is removed. Although retweets do provide redundant information, they can also be very valuable. Retweets highlight interest in a certain area. The task force can provide this information to humanitarian organizations, and further analysis can be done to determine why there is particular interest in the information being retweeted.

Verifying incoming data is an important task.[7] The Standby Task Force takes several steps to maintain quality control. The most important quality control step is performed by the verification team, which compares incoming reports with other similar reports. Typically, if the verification team can find two or three similar reports from different users, the organization considers that a more verifiable report. In addition, other tools are used to verify information, including contacting individuals with Twitter direct messages or by email if an email address is available.

Gathering information in conflict areas poses even more significant risks of misinformation and the dangers that misinformation may create, Elliott cautioned. Partnerships with those stationed in the area can help. During the Libyan crisis, the Standby Task Force was asked to support the United Nations Office for the Coordination of Humanitarian Affairs (OCHA) to provide it with on-the-ground contacts that could help in assessing the unfolding situation. Although the Standby Task Force partnered with Amnesty International for conflict mapping of Syria, the task force decided to suspend this activity until it could further examine credential and validity questions. A recurring question for the task force is whether it is possible to ensure that the information it is providing is accurate and is not putting anyone in harm's way.

More recently the Standby Task Force established the Human Resource team to monitor volunteers for burnout and to help resolve any conflicts that arise. Unfortunately, the task force has had a few instances of volunteers being disruptive. Made up of a small, devoted group whose membership is by invitation only, the human resource team works with closed communication technologies. In addition, the Standby Task Force has engaged a psychologist to look for signs of post-traumatic stress disorder (PTSD) among deployed volunteers. A few years ago, few might have felt that PTSD could be experienced by digital volunteers who may be geographically far from a disaster site. However, volunteers do suffer from the emotional and mental impacts of disaster volunteer work, and further research is needed to determine how prevalent PTSD might be, how it can be prevented, and how organizations can monitor their volunteers for it.

The Standby Task Force has also learned, reported Elliot, that it is important to continually provide feedback on how data generated by volunteers is being used. By learning how their work is benefiting others,

---

[7] For example, during the 2010 Haiti earthquake, locations would ask for resources, stating that they had no food or water, and when teams arrived at the location with supplies, they found that the location did in fact have food and water. The location was stocking supplies because there was concern about when another distribution of resources might occur.

volunteers begin to appreciate the significance of their work. This feedback is one of the primary motivation tools during a deployment.

OBSERVATIONS OF WORKSHOP PARTICIPANTS

Observations on the dynamics of social media offered by workshop panelists and participants in the discussion that followed the panel session included the following:

- Comparing the dynamics of social media use during non-emergency situations with those of emergency situations can be incredibly complicated. Each emergency situation involves unique factors that affect how social dynamics develop. For example, during a terrorist attack one can anticipate a more adversarial climate and the potential for terrorists to exploit misinformation as part of their attack, whereas during a natural disaster there is less incentive to provide misinformation. During natural disasters, misinformation typically stems from constant rereporting of old news, although there is a possibility that awareness of limited resources could create an incentive and a desire to share misinformation so as to provide oneself with supplies before others.
- An important factor in the use of social media tools and sites is the motivation of participants. As with the DARPA challenges, a financial incentive to participate can lead to a large sensor network, but can also create an inducement to interfere with others' work. If the stakes are lower, so also are the incentives to participate as well as to cause harm, thus reducing the concerns about significant interference. A question is how to use incentives to increase participation without also increasing interference. This problem is a primary reason that the Standby Task Force does not use financial incentives.
- Another option for preventing distribution of poor data is to limit the participation of anonymous workers. However, requiring that participants be non-anonymous would increase the effort required to register as a volunteer and would slow participation.
- The use of identity systems, even a readily available one such as Facebook's, also requires additional lead time, which is limited during disasters. Online identity structures are discussed further in the next chapter.
- A system that uses a hierarchy of social media users may be helpful for ensuring that information is accurate prior to its dissemination during a crisis. An example of an online hierarchy of users is Wikipedia, which provides a classic example of how increased popularity changes the dynamics of a social Web site. Wikipedia was initially very egalitarian: everyone could contribute and everyone had basic editing rights. As

Wikipedia became popular, this flat organization no longer worked, and a hierarchy of editors was created who could lock articles and exclude certain edits. But creating this sort of hierarchical system during a crisis would be quite difficult, given the time constraints of disasters and crises, which provides an incentive to be as open as possible.

4

Credibility, Authenticity, and Reputation

The use of social media to disseminate information, both official and unofficial, during disasters raises questions about how to assess the information's credibility and authenticity. For example, although the reach of an official message may be widened greatly if it is redistributed (e.g., retweeted), the message might have been modified in ways not anticipated or desired by its originators. Paul Resnick, University of Michigan; Dan Roth, University of Illinois, Urbana-Champaign; and David Stephenson, Stephenson Strategies, examined credibility, authenticity, and reputation in the context of social media and disaster response.

REPUTATION SYSTEMS

Paul Resnick observed that credibility problems have arisen in many online systems and that a variety of approaches have been explored to address them. One such approach is a *reputation system*, which formalizes the process of gathering, aggregating, and distributing information about individuals' past behavior. The electronic commerce firm eBay operates one of the largest and best-known online reputation systems, which provides buyers with a history of a seller's past transactions along with feedback from individuals who purchased items from the seller. Reputation systems have three principal functions:

- Inform participants about other participants, to help them determine if a particular participant is trustworthy.

- Create an incentive for good behavior. If participants know that they will be rated and that the rating is publicly available, they are more likely to provide accurate information (e.g., product listings), good service, and so on.
- Provide a selection effect. If participants know that good behavior will be noticed and rewarded, they are more likely to join the system. Similarly, would-be malicious participants will know that any incompetence or deliberate disruption will be made public—a deterrent to misbehavior.

Resnick cited two main challenges to reputation systems: the ability to create new pseudonyms and the high cost or other barriers to entry for newcomers. Most online sites use pseudonyms, and there are several valid reasons for not requiring "real" names.[1] A user of a reputation system who develops a bad reputation can often easily create a new pseudonym. However, reputation systems can still succeed even when users are easily able to create pseudonyms, because those that establish positive reputations will continue to use their account, thus ensuring that positive information is available in the system. Similarly, a user who establishes a positive reputation has a disincentive to suddenly shift behaviors. The lack of reputation limits that user's ability to participate in transactions since having a high approval rating with one transaction is much less valuable than having a high approval rating with 200 transactions.

Resnick also noted that the low value of having little or no reputation information creates barriers for newcomers. Research shows that it is not likely that one can treat each newcomer as having a positive reputation until they misbehave: when newcomers are treated as if they have a positive reputation, system managers become overwhelmed with the number of new, poorly behaved users that must be removed from the system. As a result, there seems to be no alternative to having newcomers pay their dues by developing a positive reputation over time. This tradeoff, between the utility of a well-managed reputation system and the high cost to newcomers, is a challenge to the growth of a reputation system, said Resnick.

Turning to the usefulness of reputation systems in the context of disaster response, Resnick commented that some participants may be able to develop a positive reputation through interactions before a disaster occurs, whereas other participants who may in fact have very useful information will not necessarily have established a prior reputation nor be able to establish their reputation quickly during an event. In such cases, additional measures are needed.

---

[1] For example, see National Research Council, *The Internet's Coming of Age*, National Academy Press, Washington, D.C., 2001.

ENCOURAGING SELF-CORRECTION

Another approach to enhancing credibility is to encourage Web users to spread correct information. When rumors of an event or disaster first appear, authorities and trusted information brokers are often tasked with broadcasting corrected information. However, rumors typically continue to spread despite having been found to be false—a problem that is not unique to the Internet or social media. Resnick mentioned that he has begun to examine the use of social processes in the context of political campaigns and elections. Traditional news media outlets have established online sites that examine the truthfulness of statements made by candidates. Although such sites are helpful to those who are already knowledgeable and are involved in politics, few others may visit these sites to verify information. One possibility is to use social media to mobilize individuals to create pointers to such correct information. A person motivated to correct false information could, for instance, begin to search tweets containing misinformation and then invite other users to respond to these incorrect tweets, including supplying links to where the information is corrected.

COMPUTATIONAL CLAIM VERIFICATION

Evaluating the trustworthiness of a particular piece of information can require connecting to many other pieces of information (either reinforcing or contradictory) from a wide variety of sources (e.g., news reports, official statements, blogs, wikis, and social media messages) that provide useful information. Metadata, such as embedded geographical coordinates or network activity associated with a piece of information, can also provide valuable information. Dan Roth observed that manually inspecting these potentially large amounts of data is difficult, a situation that prompted his research group to develop a tool to integrate relevant information to score the trustworthiness of claims and sources.[2]

Roth's research aims to create a tool that judges trustworthiness in a manner similar to how a person might. Interestingly, accuracy is not the only important factor, because information can be technically accurate yet misleading. Furthermore, simply counting the number of times information is repeated is not sufficient either. Rather, a decision on the trustworthiness of claims and sources should, according to Roth, be based on several characteristics: support for a given piece of information across multiple trusted sources, source characteristics (such as reputation), the

---

[2] V.G. Vinod Vydiswaran, ChengXiang Zhai, and Dan Roth. Content-driven trust propagation framework. *Proceedings of the 18th ACM SIGKDD International Conference on Knowledge Discovery and Data Mining* (KDD'11), 2011, pp. 974-982.

organizational type of the source (e.g., public interest, government, or commercial entity), verifiability of the provided information, and prior beliefs and background knowledge of the source.

Recognizing that a single metric based on accuracy is inadequate, Roth's research group proposed three measures of trustworthiness: truthfulness (focusing on importance-weighted accuracy); completeness (thoroughness of a collection of claims); and bias (which results from supporting a favored position with untruthful statements or targeted incompleteness/lies of omission).[3]

Roth's group's initial design was based on computing trustworthiness using sources and claims. However, this approach proved too simplistic, and other factors had to be added. These included information on the certainty (and uncertainty) of the system's technical ability to extract information, similarity across claims, attributes of group memberships, independence of sources, and so on. The system also had to incorporate prior knowledge, including some commonsense understanding and an understanding of how claims interact with one another—in order to reconcile competing truth claims, for instance.

Another aspect of checking trustworthiness is incorporating evidence. In adding another system layer for evidence, natural-language-processing techniques are needed to help determine what a message is actually saying—is it supporting or countering a specific claim?

APPLYING THE "CITIZEN SCIENCE" MODEL TO DISASTER MANAGEMENT

Past research at the University of Colorado, Boulder, and at the University of Delaware has found that during disasters individuals act largely in a self-directed, collaborative way to create emergent behavior. Their decentralized, pluralistic decision making, David Stephenson reported, finds imaginative and innovative ways to cope with the contingencies that typically appear in major disasters. Combining these emergent behaviors with social media tools could provide a significant opportunity to incorporate the public into disaster response, suggested Stephenson.

The success of citizen science initiatives, a form of crowdsourcing that harnesses individual observations to assist in scholarly research, suggests that similar techniques could be very useful in harnessing the public for help in coping with disasters. The concept is not new—the

---

[3] J. Pasternack and D. Roth. Comprehensive Trust Metrics for Information Networks. 27th Army Science Conference, November 29-December 2, 2010, Orlando, Florida.

112th Audubon Christmas Bird Count[4] was held in December 2011—but smart phones and similar technology have made it much easier to collect information.

Although the primary goal in citizen science projects is to produce research and knowledge, these projects also serve as effective outreach mechanisms, Stephenson observed. Citizen science is a powerful educational tool because it involves volunteers directly in the research process and requires the creation of simple, easy-to-follow educational programs that enable nonexpert volunteers to participate effectively.

Today, new technologies make the reporting of information much easier. For example, in a National Weather Service (NWS) program,[5] Twitter users are asked to report on weather conditions using the hashtag #WX. Through it, users can contribute information that may help the NWS better understand very localized conditions such as microbursts that may be missed by conventional instrumentation and sources. In another initiative, Tweak the Tweet,[6] a special syntax using simple, short, and easy-to-remember hashtags was developed to make Twitter messages more focused and machine readable during disasters (Figure 4.1), thus making it easier for people to contribute information in a form useful in a disaster response. The syntax was rushed into service during the 2010 Haiti earthquake recovery and helped provide needed structure to the information that residents and aid workers were reporting from the scene.

Many other possibilities are evident for the use of recent technologies in disaster reporting. Smart phones generally have precise location information, and this can be made available along with a person's messages (if a user activates this feature), making it possible for emergency managers to map the sources of reports. Another useful capability is the ability to send still images and video. Multiple pictures or video clips shot from multiple perspectives could provide authorities with a virtual comprehensive view of an event.

---

[4] The Audubon Christmas Bird Count is a census of birds performed annually by volunteer birdwatchers.

[5] See http://www.nws.noaa.gov/stormreports/ and http://www.nws.noaa.gov/stormreports/twitterStormReports_SDD.pdf.

[6] Kate Starbird, Leysia Palen, Sophia B. Liu, Sarah Vieweg, Amanda Hughes, Aaron Schram, Kenneth Mark Anderson, Mossaab Bagdouri, Joanne White, Casey McTaggart, and Chris Schenk, Promoting structured data in citizen communications during disaster response: An account of strategies for diffusion of the "Tweak the Tweet" syntax, Christine Hagar (Ed.), *Crisis Information Management: Communication and Technologies*, pp. 43-63, Chandos Publishing, 2012; K. Starbird and J. Stamberger, 2010, Tweak the tweet: Leveraging microblogging proliferation with a prescriptive grammar to support citizen reporting, short paper presented at the 7th International Information Systems for Crisis Response and Management Conference, Seattle, Wash., May 2010. Also see http://epic.cs.colorado.edu/?page_id=11.

CREDIBILITY, AUTHENTICITY, AND REPUTATION 39

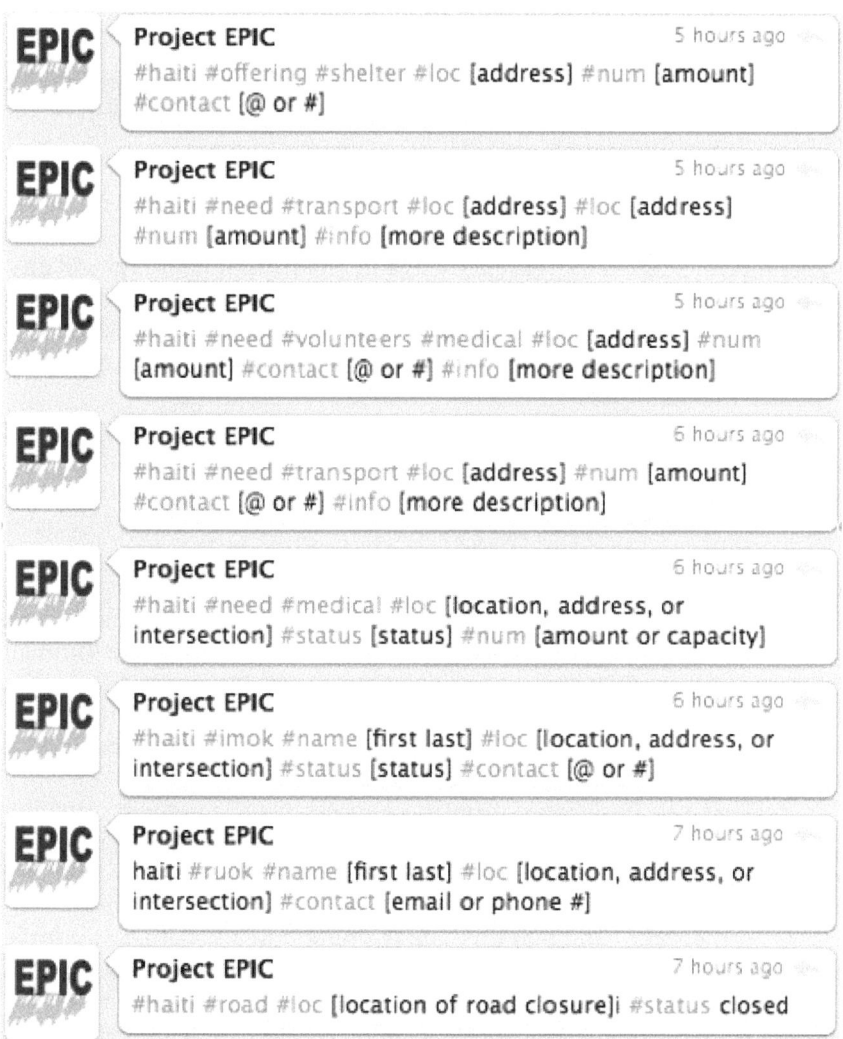

FIGURE 4.1 An example of Tweak the Tweet syntax used during the 2010 Haiti earthquake. SOURCE: Project Epic Web site. Reprinted by permission.

OBSERVATIONS OF WORKSHOP PARTICIPANTS

Observations on credibility, authenticity, and reliability offered by workshop panelists and participants in the discussion that followed the panel session included the following:

- There are many information brokers, including many not traditionally viewed as official sources of news, who can serve as trusted sources in social media, such as individuals active in a particular geographical area or in organizations such as the Standby Task Force. Many of these brokers are taking steps to verify information. Those following these trusted brokers can in turn share this information with others who may not know who is a trusted broker.
- Technology allows us to use distributed approaches to establishing trust. For example, although an individual's information may not be trustworthy, greater trust can be established if multiple reports with similar information can be found.
- The ways in which people seek information during a disaster can be different from the ways they seek information normally. For example, research has shown that during a disaster or mass emergency situation, people have a greater willingness to follow individuals who are different from themselves than they do under normal circumstances. Also, they tend to seek firsthand, "on the ground" information. Locality and hyperlocality matter. On Twitter, formal emergency response agencies or local media are retweeted more often than others.[7]
- What can be learned from past research on emergent behavior (where groups of individuals collectively complete complex tasks they could not do independently)? What prior results apply to emergent behavior with social media, and what aspects might be different?
- At the same time that they are learning how to evaluate information provided by the public, officials must also find ways to build their credibility with the public. An effort to build credibility can be as simple as an acknowledgment that an organization is listening to the public. This kind of direct communication between officials and volunteers builds a network of trust.

---

[7] Kate Starbird and Leysia Palen. Pass it on?: Retweeting in a mass emergency. *Proceedings of the Conference on Information Systems for Crisis Response and Management* (ISCRAM 2010). Seattle, Wash., May 2010.

5

Privacy and Legal Challenges with the Use of Social Media

The use of social media by emergency officials raises privacy concerns that are not present with broadcast methods of sending alerts and warnings. Official monitoring of social media to better detect or understand unfolding events is of potential value to emergency managers but may also raise privacy concerns. In addition, the networked nature of social media may provide a substantial amount of information about a single individual. For example, based on whom individuals follow on Twitter, one could infer where they live and work and where their children attend school. In addition to privacy challenges, liability and legal concerns will also have to be addressed. Peter Swire, Ohio State University, discussed legal and policy perspectives on privacy challenges. Bryan Ware, Digital Sandbox, discussed the privacy challenges that limit tool use and how privacy protections might be built into application development. Aram Dobalian, VHA Emergency Management Evaluation Center, explored liability issues related to first-responder use of social media.

LEGAL AND POLICY PERSPECTIVES ON PRIVACY AND ON GOVERNMENT MONITORING OF SOCIAL MEDIA

Swire described two main points of view regarding the government's monitoring of social media to gain situational awareness during an emergency. One perspective is that because individuals have posted information online and made it readily available to the public, the government should feel comfortable using that information to gain situational awareness and respond to needs during a crisis. The second perspective is that many people find it disconcerting to know that the government is reading and analyzing information posted in social media.[1] The monitoring raises several questions: How long will such information be stored? What else will it be used for? Will individual dossiers be created that may potentially lead to limitations on an individual's freedoms?

Swire observed that the Fourth Amendment to the U.S. Constitution, with its limits on search and seizure by the government, helps shape understanding of what is public. Essentially, probable cause or a warrant is needed to enter an individual's house or vehicle.

By contrast, the Fourth Amendment does not limit the government's ability to follow people on a public street or to read information published in a newspaper, a precedent that might be extended to cover government monitoring of public social media communications. Interestingly, in the context of another circumstance involving new technology and its privacy implications, the January 2012 Supreme Court decision in *United States v. Jones* placed limits on the definition of "in public." The court issued a unanimous opinion that a warrant was needed to place a GPS tracking device on an automobile even though the vehicle in question was traveling in public spaces. The majority of justices emphasized that physically attaching something to a car was a factor in the decision; other justices questioned whether "in public" is enough to make surveillance acceptable.

Another Fourth Amendment consideration relevant to social media is consent—individuals can consent to a search or seizure. When people make information available to the public through social media, does this action mean that consent has been given? A related question is under what circumstances people give consent. Facebook user settings provide an example of the lack of clarity surrounding the concept of consent. Several papers have discussed how Facebook users have trouble understanding their privacy settings.[2] Users often believe that postings are not available to the public when in fact they might be.

---

[1] Concerns about government use of information can, observed Swire, be attributed at least in part to past abuses. For example, it was uncovered after the fact that the FBI had placed many of the delegates at the 1972 Democratic party national convention under surveillance. In the wake of Watergate and other abuses, President Ford's attorney general issued what would become known as the Levi guidelines to limit the information, including public information, that law enforcement could gather. Concerns about government intrusion also led to passage of the 1974 Privacy Act.

[2] Maritza Johnson, Serge Egelman, and Steven M. Bellovin. Facebook and privacy: It's complicated. Symposium on Usable Privacy and Security (SOUPS), July 2012. Michelle Madejski, Maritza Johnson, and Steven M. Bellovin. A study of privacy setting errors in an online social network. *Proceedings of SESOC 2012*, 2012. An earlier version is available as Technical Report CUCS-010-11.

Swire also discussed several areas of federal legislation and their possible implications for government monitoring of social media:

- The 1974 Privacy Act regulates how a government system of records is maintained and how information is shared with other federal agencies and individuals. Pursuant to this act, the Department of Homeland Security (DHS) released a System of Records Notice in February 2011 regarding its monitoring of social media.[3] Provided that a set of procedures outlined in the regulation is followed, there is no strict constraint on data collection and use.
- State and federal wiretapping statutes place strict limits on government interception of telephone calls. While this constraint does not apply directly to items posted using social media, it highlights how protective the law can be regarding the monitoring of communications that are intended to be private.
- The Stored Communications Act applies to records held by a third party. Although a subpoena to the third party is required, there is a medium level of strictness around the collection of this information.
- Legislation regarding so-called pen registers concerns the collection of pen register information—who an individual calls, texts, or emails—but not the content of the communication. Although a judicial order is needed, the standard for receiving an order is fairly low. Routing information has not been considered particularly sensitive but is fairly useful to law enforcement. In addition, friends' lists on Facebook are by default public.

The state of the law regarding government access to and use of location information available from mobile devices or social media is still being examined in the courts. The lower courts are currently split on whether a warrant is needed to track an individual's cellular phone. This is a particularly sensitive area because cellular devices can track the location, time, and date of a person's activities, potentially revealing a broad picture of an individual's life.

Finally, Swire observed, although certain surveillance or monitoring activities may be legal, they may not be desirable as a matter of government policy. Indeed, the Department of Homeland Security (DHS) has recognized this concern and has itself imposed limitations on its monitoring activities. DHS monitors some public officials and traditional, new, and social media; the restriction is that individuals are not tracked. These

---

[3] Department of Homeland Security. Publicly Available Social Media Monitoring and Situational Awareness Initiative System of Records. FR Doc. No. 2011-2198, 2011. Available at http://www.gpo.gov/fdsys/pkg/FR-2011-02-01/html/2011-2198.htm.

limitations were included in DHS's System of Records Notice, which makes these rules binding. Nevertheless, at a recent hearing in the House of Representatives, Congress criticized DHS's approach to social media.[4]

PRIVACY PROTECTION IN THE CONTEXT OF PROGRAMS FOR CITIZEN REPORTING OF THREATS

Bryan Ware described the "See Something, Say Something" (S4) campaign and other efforts to encourage the public to report potential threats, and used them to illustrate how privacy considerations can constrain the development and deployment of applications and tools that enable the public to provide information on potential threats.

The S4 program, which is run out of DHS's Office of Public Affairs, is intended primarily to increase the public's awareness of safety issues, and only secondarily to collect information. Accordingly, it does not provide a mechanism for making reports, other than the 911 emergency phone number that is ubiquitous in the United States. Each public safety organization has a unique reporting phone number or other reporting mechanism.

Although there would seem to be considerable scope for accepting reports from smart phone applications and other forms of social media, Ware observed that their adoption was being held back in part by concern about privacy and civil liberties. For example, Ware's firm, Digital Sandbox, Inc., developed an S4 reporting tool for mobile devices that was never deployed owing to unresolved privacy issues. Although the S4 program was thus ultimately not successful, the experience provides valuable lessons for future endeavors.

Another system, Threat and Risk Incident Management, also built by Digital Sandbox, Inc., was developed to provide capabilities similar to those of the S4 application and was deployed during the 2012 Super Bowl. The system, built to be used during the 10 days of Super Bowl activities, combined several information sources—911 calls, event activities and locations, field reports from trained personal, and very limited monitoring of social media. Although the system could have been used to collect citizen reports (such as pictures taken with a smart phone), its use was limited to trained personnel, again because of privacy concerns. Monitoring of social media was limited; for example, Twitter feeds were searched only for certain terms in certain areas, and Twitter users' names and associated URLs were not collected. Despite these constraints, Ware

---

[4] Subcommittee on Counterterrorism and Intelligence. DHS Monitoring of Social Networking and Media: Enhancing Intelligence Gathering and Ensuring Privacy. February 16, 2012. Washington, D.C. Available at http://homeland.house.gov/hearing/subcommittee-hearing-dhs-monitoring-social-networking-and-media-enhancing-intelligence.

said that the tool proved very useful because it provided emergency managers an integrated view of potential events and also made this information available to officials on their mobile devices.

LEGAL PERSPECTIVE ON FIRST-RESPONDER RESPONSIBILITIES

Aram Dobalian explored the use of social media for emergency response, including both informal and formal and paid and volunteer responders, focusing first on a set of legal issues related to the use of social media for emergency response. He observed that the case law in this area is very limited, but that there are some relevant legal perspectives that can be applied, including the tort of negligence, duty of care, and Good Samaritan laws:

- *Tort of negligence.* There are four basic elements to the tort of negligence: duty, breach, causation, and damages. These form the basis, for example, for most medical malpractice claims and might be most relevant to first responders and social media use. Negligence states that "a failure to provide the care that a reasonably prudent responder would provide under the same or similar circumstances might be the basis for liability." For negligence to be found, there first must be a certain duty or responsibility that is owed by the responder to the victim. Second, that duty has to be breached in some form. Third, as a result of that breach, the victim must be further injured or harmed.
- *Duty of care.* In most cases, bystanders have no duty or responsibility to provide aid to an injured individual.[5] However, courts and statutes have created several exceptions. One is that if someone begins to provide aid to an individual in distress, that person must continue to provide aid at least until someone more qualified can provide the aid—the rationale for the requirement being that once someone starts, others are less likely to step in to provide assistance. In the context of social media, this stance suggests that a first responder who reads a social media post requesting assistance incurs obligations once he or she communicates an intent to provide assistance.

A second exception is that if someone creates a situation that causes harm, that person has the duty to mitigate damage caused. In the context of social media, messages that, for example, led to a stampede could create liability.

A third exception is when the rescuer has a caretaker or other special

---

[5] In two states, Vermont and Minnesota, laws do require that assistance be provided. Vermont imposes a fine, and Minnesota treats failure to provide aid as a misdemeanor, although the obligation can typically be met by calling 911.

relationship (including teacher-student, business-customer, and provider-patient) to the victim. A special relationship may also be created if the rescuer starts to provide advice on, say, how to treat injuries. However, there is very little case law that is relevant in this particular area. There may be some similarities as well to teledelivery of health services, but again there is still very little case law regarding remote diagnosis and treatment.

- *Good Samaritan laws.* These laws shield individuals from liability when they offer assistance provided that they act rationally, in good faith, and in accord with whatever relevant training they may have—and when their assistance is not being provided in conjunction with their employment. The intent of such laws is to encourage bystanders to provide assistance without fear of incurring liability. Jurisdictions vary in their application of these laws to people with medical or other specialized training, and some apply only when victims are in imminent danger. Also, the laws do not apply to those who work in an emergency response-related profession if they are being paid at the time to provide aid. When they are acting as volunteers, how the laws apply is less clear-cut.

Because social media interactions readily span jurisdictions, their use in disasters raises issues that also arise with telemedicine. There are, for example, unresolved questions about the circumstances under which health care professionals can provide advice or care to someone located in a state where the professional is not licensed. There are also questions about what the standards of care should be when advice is provided remotely, without an opportunity for a physical examination.

Social media interactions also raise privacy questions because information transmitted over social media may be public or semipublic, yet a health care provider may have obligations under the Health Insurance Portability and Accountability Act (HIPAA) or other privacy laws[6] that are difficult to meet under those circumstances.

Given all of these challenges, it is perhaps not a surprise that emergency managers have not been able to put formal systems in place that use social media for emergency response, commented Dobalian. However, there is at least one example of such a service—the United Hatzalah in Israel, which coordinates a group of approximately 1,700 volunteers who are trained in first aid and have GPS-enabled smart phones. A smart phone application notifies the nearest responder to attend to an event—with a goal of responding within 90 seconds to a Twitter message or

---

[6] Department of Health and Human Services. Hurricane Katrina Bulletin: HIPAA Privacy and Disclosures in Emergency Situations. 2005. Also see "Disclosures in Emergency Situations" under Frequently Asked Questions at http://www.hhs.gov/ocr/privacy/hipaa/faq/disclosures_in_emergency_situations/index.html.

other post signaling the event, well before an ambulance normally would arrive.[7] If the United States were to create an analogous volunteer-based system, the legal issues outlined above would have to be considered, observed Dobalian. And, Dobalian commented, if social media were to be used for professional emergency response, a number of questions would have to be addressed, such as the following:

- How would requests via social media be validated?
- How would the architecture for such a system resemble and differ from existing 911 telephone systems?
- What levels of staffing and skills would be needed to effectively monitor social media for requests for assistance?
- How would the information available to dispatchers be different from that available to a 911 operator, and what would the implications be for what resources were dispatched and what advance instructions were given?

OBSERVATIONS OF WORKSHOP PARTICIPANTS

Observations on privacy and legal issues associated with social media offered by workshop panelists and participants in the discussion that followed the panel session included the following:

- Messages that do not contain a verified identity will often suffice as a source of information when events are being monitored or the public's response is being assessed, and people may be more comfortable remaining anonymous when their communications are monitored by the government. However, people's attempts to remain anonymous are complicated because a user's network may provide clues to his or her identity.
- The distinction between personal identifiable information and other information about people is rapidly eroding as researchers have come to understand that fairly innocuous attributes can pinpoint individuals. Are there other techniques or approaches that can be used to disguise a person's identity?
- There is a distinction between impersonal trust (trust in institutions) and interpersonal trust (trust in other individuals). During a crisis, trust has interesting dynamics—both impersonal and interpersonal trust can be eroded, or one may serve as a substitute for the other. How are these dynamics altered by the use of social media during a disaster?
- The U.S. Department of Health and Human Services released new regulations regarding medical privacy in the aftermath of Hurricane

---

[7] See http://www.hatzalahrl.org/index.php.

Katrina that allow health care providers to better share patient information as necessary to provide treatment in an emergency. Health care providers can also share patient information as necessary to identify, locate, and notify family members, guardians, or anyone else responsible for an individual's care. Could these guidelines serve as a more general model for how to relax privacy restrictions during emergencies? Could new HIPAA guidelines regarding de-identified health data provide a helpful starting point?

- Copyright issues, especially with respect to photographs, may create an additional challenge to the use of social media, although copyright protects expression rather than the underlying data or idea. If photographs were used simply to identify patterns, this use of data would not create a copyright problem.

6

Research Gaps and Implementation Challenges

Social media represent a relatively new and still rapidly evolving phenomenon, and their application to alerts, warnings, and other aspects of emergency management is still in its infancy. But there is much current interest in the use of social media because they have been embraced by a large segment of the population and because they enable new, two-way interactions among those affected by and responding to disasters. To date, formal study of the use of social media in disasters has been limited (Box 6.1 explores the state of research on the use of social media in emergency management), and there are many outstanding questions about how they can be used most effectively by emergency managers and other public officials, organizations, communities, and individuals.

The following sections outline research opportunities and associated implementation challenges identified by the committee and workshop attendees during the plenary and breakout sessions of the workshop. The opportunities and challenges compiled here from presentations and discussions at the workshop do not reflect a consensus of the committee or the workshop participants, nor are they intended to be a comprehensive list of research questions.

MESSAGE CONTENT AND DISSEMINATION

A significant body of past research has considered what types of messages and communications strategies are most effective for alerting the public with traditional emergency alerting tools like broadcast radio

> **BOX 6.1**
> **State of Research on Social Media in Emergency Management**
>
> In her remarks at the February 2012 workshop on alerts and warnings using social media, Leysia Palen of the University of Colorado, Boulder, discussed the evolving application of social media for emergency management and the associated stages of research maturity. She suggested that growing interest in examining the role of social media reflects in part the progress that has been made toward their adoption, and that research together with learning from the practical application of social media will increase understanding of both possibilities and pitfalls and thus foster greater, more effective use.
>
> From roughly 2008 to 2011 was a period in which the potential for using social media was first recognized and was marked by scattered grassroots experimentation, said Palen. In this first stage, publications by practitioners and researchers, workshops, and discussion developed a case that social media would inevitably play an important role in emergency management, although just how was unclear. Not all embraced the new technologies. Some felt that the use of social media was simply a passing fad, and even as late as 2011 otherwise knowledgeable people remained fearful about social-media-abetted change and sought to understand how social media could be "held back." Still others embraced the trend but did not fully understand its grassroots and spontaneous nature; one result was attempts to shape it in order to gain commercial or tactical advantage.
>
> Indications abound that both practice and research have since yielded significant advances, observed Palen. Local emergency managers are experimenting with how to incorporate social media into their daily practices, for example, and the American Red Cross has incorporated certified volunteers into its social media response plans. Formal policy discussions are being held worldwide.

and television.[1] Comparatively little research has examined similar questions for messages disseminated via social media.[2] One of social media's

---

[1] National Research Council. *Public Response to Alerts and Warnings on Mobile Devices: Summary of a Workshop on Current Knowledge and Research Gaps.* The National Academies Press, Washington, D.C., 2011.

[2] Research that has been done in this area includes Kate Starbird, Leysia Palen, Amanda Hughes, and Sarah Vieweg, Chatter on the red: What hazards threat reveals about the social life of microblogged information, *Proceedings of the ACM 2010 Conference on Computer Supported Cooperative Work* (CSCW 2010), pp. 241-250, 2010; Kate Starbird and Leysia Palen, Pass it on?: Retweeting in mass emergencies, *Proceedings of the Conference on Information Systems for Crisis Response and Management* (ISCRAM 2010), Seattle, Wash., 2010; Leysia Palen, Sarah Vieweg, Sophia Liu, and Amanda Hughes, Crisis in a networked world: Features of computer-mediated communication in the April 16, 2007, Virginia Tech event, *Social Science Computing Review,* Sage, pp. 467-480, 2009; and Clarence Wardell and Yee San Su, *Social Media + Emergency Management Camp: Transforming the Response Enterprise,* 2011, available at http://www.wilsoncenter.org/sites/default/files/SMEM_Report.pdf.

particular strengths, that messages can be widely shared, also presents challenges because messages can be readily altered as they are spread. In addition, messages that may no longer be accurate can continue to propagate through social media long after they are no longer current. Their interactive nature makes social media useful as a medium for both receiving and confirming disaster information, which suggests opportunities to reduce the gap between the time individuals receive disaster information and when they take action.

Some specific research questions include the following:

- How should broadcast messages from emergency managers be crafted in light of the limitations (e.g., short message lengths) and strengths (e.g., opportunities to include images, maps, and URLs) presented by social media?
- How much of the word-of-mouth dissemination of information about disasters occurs through social media? Are there ways of designing messages that could increase the speed and breadth of their spread?
- How are messages altered as they are spread through social networks? How might messages be formulated to discourage or reduce the impact of these changes?
- What strategies and techniques can be applied to deal with messages that have "aged" to the point that they are no longer relevant?
- What types of messages and strategies would reduce the time lag before individuals take action (i.e., reduce milling time)?
- What challenges or opportunities will social media present in reaching unique populations such as non-English speakers or individuals with disabilities?

TRUST AND CREDIBILITY

In addition to sharing and commenting on messages they receive, citizens often use social media to share firsthand text, image, and video reports about disasters. This firsthand information can be useful for decision making by emergency officials as well as other individuals but raises questions about how to assess its trustworthiness. The nature of social media suggests the possibility for self-correcting information by combining reports supplied by many individuals provided that the number of reports is sufficiently large.

Some specific research questions include the following:

- How do consumers of social media messages distinguish credible from less credible information? How can emergency managers and other officials create and disseminate messages that have high credibility?

- What are practical ways that officials can evaluate and signal the credibility of unofficial messages during an event?
- What are the relationships between the number and density of social media users or the size of an event and the effectiveness of mechanisms for self-correcting information users supply?
- What mechanisms and approaches foster such self-correction?
- What are effective strategies officials can use to intercede when misinformation is proliferated via social media?

PRIVACY

The use of social media for alerts and warnings raises privacy issues that were not in play with traditional methods of sending alerts and warnings. For example, the social media communications being monitored by government officials, while technically public, may have been sent with certain expectations of privacy such as that they would not be read by government officials.

Some specific research questions include the following:

- How, if at all, do people differentiate the privacy implications of message monitoring by government agencies, by commercial entities, and by the general public during disasters versus at other times?
- It has been suggested that people are willing to accept reduced privacy safeguards during disasters. What are people's actual attitudes in these circumstances?
- How might the government's use of social media be adjusted during disasters? For example, are there mechanisms that could be used to trigger monitoring when a disaster begins? What safeguards could be established to ensure that people have full control of adjustment to and reactivation of privacy settings?
- Is widespread adoption of social media, which relies on users sharing information about themselves, altering the privacy expectations of users of social media? What are the implications for the use of social media during disasters?

VOLUNTEERS

Social media have enabled the emergence of online groups of volunteers to respond to disasters. Some of these groups, such as the Standby Task Force and Humanity Road, have evolved from ad hoc groups to more structured volunteer organizations that designate individuals responsible for coordinating response activities. These more formal groups as well as spontaneously formed groups help curate disaster information from social media and other sources and use social media to provide relevant information to both official responders and the affected population.

Some specific research questions include the following:

- What organizational theory provides an understanding of how ad hoc volunteer organizations form, function, and evolve—and what the implications are for disaster management?
- Although official first responders in government and nongovernment organizations have had training to deal with emergency situations, most ad hoc volunteers have not. Are there ways that social media can be used to make the efforts of ad hoc volunteers more effective?
- How do legal and policy concerns constrain the interactions of volunteers with formal emergency managers? What measures might be taken to address these concerns?
- What are points of cooperation and tension between officials and volunteers?

TECHNOLOGY DIFFUSION

Several instances of technologies that could have immediate application for disaster management were discussed during the workshop, such as support for visualization of information derived from social media. However, it was also evident that there were relatively few points of engagement between researchers developing or investigating new tools and emergency managers and other potential end users. Emergency managers are most likely to encounter new technologies only when such tools are made available by vendors. Given the rapid pace of change in social media and the associated rapid pace of change in tools for using social media, workshop participants suggested that more rapid and effective technology transfer would be valuable.

Some specific research questions include the following:

- Are there emerging best practices for how social media can be used effectively by emergency managers?
- How can diffusion of available technologies be promoted? What are the special characteristics of the emergency management community that limit the adoption of new technologies and techniques, and how might such characteristics be addressed?
- How can the growing body of knowledge on how users behave in online communities be transferred to emergency management practices?

EMERGENCY MANAGEMENT PRACTICE

Although the rate of adoption of social media and the sophistication of their use by emergency managers vary considerably, it does appear that emergency managers have come to generally appreciate the potential value of social media. However, workshop participants cited a number of barriers that still exist to the effective use of social media in the practice of emergency management. These barriers stem in no small part from an incomplete understanding, as discussed above, of how to use social media in disasters and the relative newness of the medium.

Some specific challenges include the following:

- Limited knowledge about information-sharing techniques and collection of information;
- Limited staff plus budget challenges that create barriers to using social media for situational awareness;
- Lack of policies and discussion about the use of social media for dissemination of alerts and warnings and for situational awareness; and
- Concerns about potential liabilities that are created when new technology is introduced, specifically with respect to fair representation of victims' needs and the distribution of resources.

Appendixes

A

Workshop Agenda

**FEBRUARY 28-29, 2012
BECKMAN CENTER
IRVINE, CALIFORNIA**

Tuesday, February 28, 2012

8:30 am Welcome and Opening Comments

*Robert Kraut, Chair, Committee on Public Response to Alerts and Warnings Using Social Media
Denis Gusty, Department of Homeland Security*

9:00 Fundamentals of Alerts, Warnings, and Social Media

Much is known about the public response to alerts delivered by sirens, radio, television, and weather radio. As social media play an increasingly important role in societal communication, it will become increasingly important to understand the implications of these new capabilities for disaster alerts and warnings.

What is known about how the public responds to alerts and warnings?
Dennis Mileti, University of Colorado, Boulder

What is known about the use of social media during a disaster?
Kristiana Almeida, American Red Cross

What are barriers to official use of social media during a disaster?

Edward Hopkins, Maryland State Emergency Management Agency

What technologies are in development for alert dissemination and situational awareness via social media?
Emre Gunduzhan, Johns Hopkins University Applied Physics Laboratory

Timothy Sellnow, University of Kentucky, moderator

10:30 **Dynamics of Social Media**

The social aspect of these tools makes them especially attractive because of the ability to leverage the trust people place in their connections. Information about an event that is provided by neighbors, colleagues, friends, or family is often viewed as more credible than a mass alert or a news report. Social media may also provide a useful complement to other tools by providing a way to rapidly disseminate time-sensitive information that may be important to an affected community but not rise to the level of an official alert or warning. How connections form, how information is disseminated, and why users volunteer their time and knowledge to solve problems have been examined by researchers in human-computer interaction, psychology, and computer science. The panel will explore what motivates people to participate in knowledge sharing, what drives self-organizing, and what mechanisms exist for self-correction of information.

Influence mechanisms in social media
Duncan Watts, Yahoo! Research

Incentivizing participation in time-critical situations
Manuel Cebrian, University of California, San Diego

How the Standby Task Force harnesses the power of the crowd
Melissa Elliott, Standby Task Force

Jon Kleinberg, Cornell University, moderator

Noon Lunch

APPENDIX A

1:00 pm   **Credibility, Authenticity, and Reputation**

During disasters, citizens often post firsthand information and pictures and re-post information they have received from official or unofficial sources. Although both types of information are useful to both emergency officials and the public, such sharing raises questions about how to assess the credibility and authenticity of firsthand reports and redistributed information. For example, although the reach of an official message may be widened if it is redistributed (e.g., retweeted), the message may have been modified in ways not anticipated or desired by its originators. The panel will explore credibility, authenticity and reputation in the context of social media and disasters.

Information verification and rumor control
Paul Resnick, University of Michigan

Mechanisms for determining trustworthiness
Dan Roth, University of Illinois, Urbana-Champaign

Training the public to provide useful data during a disaster
David Stephenson, Stephenson Strategies, Medfield, Mass.

Leysia Palen, University of Colorado, Boulder, moderator

2:30   **Personal Privacy**

The use of social media by emergency officials raises privacy concerns that were not present with traditional methods of sending alerts and warnings. Also privacy-sensitive, but of potential value to emergency managers, is official monitoring of social media to better detect or understand unfolding events. For example, the networked nature of social media may provide a substantial amount of information about a single individual: based on who one follows on Twitter one may be able to infer where she lives or works and what school her children attend. The panel will consider such questions as:

- What are the public's perceptions and expectations of privacy, and how can they best be addressed? For

example, the communications being monitored by government officials, while technically public, may have been sent with certain expectations of privacy such as not being intended to be read by government officials.
- What is the appropriate balance of interests between achieving effective situational awareness and privacy? For example, how should location-tagged information be handled?
- What are best practices in providing adequate notice to the public and ensuring that collected information is used appropriately? For example, how can or should users whose public information is being monitored be made aware of that? How frequently should notice be provided?
- Are there existing features of social media that could be used to help protect privacy? For example, would asking people to use designated mechanisms (e.g., hash tags in Twitter) to label information they intend to be read by government officials constitute an adequate opt-in approach?

Privacy decision making
Lorrie Cranor, Carnegie Mellon University

Social-psychological challenges of social media use in crises
Gloria Mark, University of California, Irvine

Implementation of the "See Something, Say Something" campaign—how privacy can be protected
Bryan Ware, Digital Sandbox

Today's framework for privacy protection and its application to alerts and warnings using social media
Peter Swire, Moritz College of Law, Ohio State University (remotely)

Alessandro Acquisti, Carnegie Mellon University, moderator

4:00	Break
4:15	**Breakout discussion** on opportunities and challenges.
5:30 pm	Reception

APPENDIX A

Wednesday, February 29, 2012

8:30 am	Report-backs from breakout sessions
9:30	**Case Studies of Uses of Social Media in Disasters**

Social media is already being used both formally and informally by emergency managers. Researchers have also begun to examine social media communication streams to learn how social media are used during a disaster. This panel will examine recent experience and research on social media use.

Currently used tools for monitoring social media for situational awareness
Brian Humphrey, Los Angeles Fire Department

Use of Twitter for earthquake detection and alerting
Paul Earle, USGS National Earthquake Information Center

The use of social media tools to disseminate information during a health crisis
Keri Lubell, Centers for Disease Control and Prevention

Leslie Luke, Office of Emergency Services, County of San Diego and *Richard Muth, Maryland Emergency Management Agency*, moderators

10:30   **Use of Social Media by Nongovernment Organizations**

News organizations and technology firms have used social media during crises and disasters to provide information to and gather information from the public. This panel will explore lessons for government from this private-sector experience, partnerships between the public and private sectors, and how new technology may shape those partnerships.

Brad Panovich, News Channel 36, Charlotte, North Carolina

Robert Kraut, Carnegie Mellon University, moderator

11:15   **Looking Ahead: Opportunities and Challenges**

What changes in preparation, management, and analysis will be needed to incorporate social media as an information tool?

Murray Turoff, New Jersey Institute of Technology (remotely)

Social media: Legal perspectives on first-responder responsibilities
Aram Dobalian VHA Emergency Management Evaluation Center

Spontaneous and organized digital volunteerism in the future of emergency management
Leysia Palen, University of Colorado, Boulder

Michele Wood, California State University, Fullerton, moderator

12:30 pm   **Wrap-up Panel and Plenary Discussion**

Denis Gusty, DHS
Robert Kraut, Carnegie Mellon University
Leysia Palen, University of Colorado, Boulder

1:00 pm   Adjourn/Lunch

B

Biosketches of Workshop Speakers

Kristiana Almeida has been with the American Red Cross for more than 4 years. Most recently, she has been serving as a consultant to the organization's more than 250 socially active chapters, helping them to reach the next level of online engagement with their constituents while engaging new volunteers to help with social programs. She also serves as part of the organization's Advanced Public Affairs Team and as such has been a key on-the-ground media spokesperson after large-scale disasters, including the tornadoes in Alabama, the historic flooding in Tennessee, and Hurricane Irene. Almeida received her bachelor's degree from the University of California, Santa Barbara, where she graduated with distinction in her major, and she is currently working on her MBA from the University of Massachusetts, Amherst.

Manuel Cebrian works at the intersection of the computer and social sciences. He is currently an assistant research scientist with the Department of Computer Science and Engineering at the University of California, San Diego. Prior to joining UC San Diego, Cebrian was a Fulbright Fellow with the MIT Media Laboratory. Previously, Cebrian worked with Facebook, Telefonica, and Brown University. Cebrian earned a PhD in computer science from the Autonomous University of Madrid, Spain.

Lorrie Faith Cranor is an associate professor of computer science and of engineering and public policy at Carnegie Mellon University, where she is director of the CyLab Usable Privacy and Security Laboratory

(CUPS). She is also a co-founder of Wombat Security Technologies, Inc. She has authored more than 100 research papers on online privacy, usable security, phishing, spam, electronic voting, anonymous publishing, and other topics. She has played a key role in building the usable privacy and security research community, having co-edited the seminal book *Security and Usability* (O'Reilly, 2005) and founded the Symposium on Usable Privacy and Security (SOUPS). She also chaired the Platform for Privacy Preferences Project (P3P) Specification Working Group at the W3C and authored the book *Web Privacy with P3P* (O'Reilly, 2002). She has served on a number of boards, including the Electronic Frontier Foundation board of directors, and on the editorial boards of several journals. In 2003 she was named one of the top 100 innovators 35 or younger by *Technology Review* magazine. She was previously a researcher at AT&T-Labs Research and taught in the Stern School of Business at New York University.

Aram Dobalian is the director of the Department of Veterans Affairs' Emergency Management Evaluation Center (VEMEC) at the VA Greater Los Angeles Healthcare System (VAGLAHS). VEMEC's mission is to promote the health and social welfare of veterans and the nation before, during, and after national emergencies and disasters through research and evaluation. Dobalian is also an associate adjunct professor of health services at the Jonathan and Karin Fielding School of Public Health at the University of California, Los Angeles (UCLA). Dobalian received his PhD in health services from the UCLA School of Public Health with an academic cognate in social psychology, and his JD from Whittier Law School where he was editor-in-chief of the *Whittier Law Review*. He received his MPH in health services from UCLA and his BS in physics from Vanderbilt University. From 2001 to 2004, Dobalian was an assistant professor in the Department of Health Services Research, Management and Policy at the University of Florida. His research focuses on emergency management/public health emergency preparedness and response, including the impact of bioterrorism, hurricanes, earthquakes, and other natural and human-caused emergencies and disasters. His research also spans nursing, long-term care, nursing home malpractice, advance care planning, and the role of pain in the use of health services. In 2009 to 2010, Dobalian led the development of the first national VA Comprehensive Emergency Management Program Evaluation and Research agenda. The goals of this agenda were to provide a basis for fostering the conduct of VA-based emergency management research, to promote new discoveries and improve care delivery during and after emergencies, and to position VA as a national leader in emergency management research.

Paul Earle currently serves as the director of operations of the USGS National Earthquake Information Center (NEIC). His primary responsibility is oversight of 24/7 earthquake monitoring. In this capacity, he guides the development and implementation of new policy and procedures used during earthquake response and catalog production. He also serves in the rotating role of NEIC event coordinator, overseeing the production of near-real-time products following earthquake disasters around the globe. Previously, Earle graduated from the University of California, Berkeley, with a BA in geophysics and then received a PhD in geophysics from the Scripps Institution of Oceanography at the University of California, San Diego, and had a National Science Foundation postdoctoral fellowship at the University of California, Los Angeles. His research has included studies of the fine-scale structure of the deep Earth, characterization of Earth's seismic signals, and post-earthquake impact assessment as part the the USGS Prompt Assessment of Global Earthquakes for Response (PAGER) project.

Melissa Elliott works to bridge the gap between technology and teamwork. She is a core team member of the Standby Task Force and has actively participated in deployments for Al-jazeera, Amnesty International USA, OCHA, UNHRC, UN-Spider, and WHO. An avid supporter of the Haitian relief efforts, Elliott was an early adopter of the Ushahidi platform to coordinate aid after the 2010 Haiti earthquake and has traveled to Haiti on multiple occasions to assist. She is a member of Crisis-Commons and CrisisMappers. She has presented social media techniques used in disaster relief to the American Red Cross during its Emergency Social Data Summit in Washington, D.C., as well as multiple presentations at the Canadian government's Department of Foreign Trade and International Affairs during its Open Innovation Summit in Ottawa. Elliot is also a partner and executive producer at Blackbox Communications in Toronto.

Emre Gunduzhan received his BS and MS degrees in electrical engineering from Bilkent University, Turkey, and his PhD degree, also in electrical engineering, from the University of Maryland, College Park. He worked at the Advanced Technology Research Group of Nortel for more than 10 years before joining the Johns Hopkins University Applied Physics Laboratory, where he is currently a project manager in the Communication Systems Group. His current interests are modeling, analysis, and design of complex communication systems, including alert and warning systems, wireless networks, and satellite networks.

Edward Hopkins serves as director of external affairs and communications for the Maryland Emergency Management Agency (MEMA), where he oversees all external communications and liaisons, messaging, and legislative affairs. He previously served as chief of staff, deputy director of operations, and manager of MEMA's Office of Domestic Preparedness–Law Enforcement Liaison Group, where he served as the law enforcement/intelligence liaison with all state, local, and federal law-enforcement agencies and oversaw DHS grants totaling more than $10 million. Hopkins previously served as the director of communications with Maryland's Department of Juvenile Services, where he was responsible for external communications with the statewide news media, and developing press conferences, presentations, speeches, and talking points on behalf of former Governor Robert Ehrlich. Hopkins served with the Harford County (MD) Sheriff's Office for 29 years and retired in 2003 as a lieutenant. From 1994 to 2002 and from 2003 to 2005 he served as director of public and media relations and chief spokesman for the Sheriff's Office. During his career, Hopkins served as director of the Harford County Child Advocacy Center, a center for the investigation of child sexual abuse, and for 11 years was assigned to the Criminal Investigation Division, where he was a supervisor of the Major Crimes Unit. Hopkins holds a master's degree in public administration with a minor in police management from the University of Baltimore. He has attended numerous law enforcement seminars and has participated in panel discussions regarding the role of law enforcement in public affairs. He formerly hosted and produced a cable television show entitled "Behind the Badge," an award-winning informational show about the Harford County Sheriff's Office, and currently hosts "Inside Harford County," a live call-in talk show on the Harford Cable Network. Hopkins is a graduate of the FBI/Harford County Sheriff's Office Law Enforcement Executive Development School and has served as president and vice-president of Harvard Associates in Police Science, a 600-member medico-legal death investigation educational organization. He is actively involved in the community, having served 38 years with the Bel Air Volunteer Fire Company and serving currently as fire chief. He has held the positions of board president and assistant fire chief, paramedic, and public information officer. He currently serves as chair of the Maryland Municipal League's Hometown Emergency Preparedness Ad Hoc Committee, a subcommittee whose mission is to educate municipalities on emergency preparedness, planning, and awareness. In 2011 Hopkins was re-elected to local political office and currently serves as mayor of the Town of Bel Air, Maryland.

Brian Humphrey joined the Los Angeles Fire Department in 1985, where he has served with distinction, earning honors on the battle lines of countless storms, conflagrations, and disasters—including the Los Angeles riots of 1992 that caused more than $1 billion in property damage and left nearly six dozen dead. During the past 17 years, he worked fulltime in external relations for the Los Angeles Fire Department, dealing firsthand with all aspects of print, radio, television, and Internet journalism. Humphrey is active in a variety of government and diplomatic affairs, including service as an LAFD terrorism liaison officer, public safety ambassador for visiting dignitaries, and manager of the popular LAFD news and information blog.

Keri M. Lubell, PhD, is the lead for the Communication Surveillance and Evaluation Team and acting lead for the Emergency Web and Social Media Team in the Emergency Risk Communication (ECS) Branch, Division of Emergency Operations, at the Centers for Disease Control and Prevention. Her current work focuses on developing efficient and effective systems for gathering information from and analyzing the communication environment (news and social media) during emergencies in order to inform agency communication strategy. She also coordinates several efforts to evaluate CDC communication outreach activities during emergency health threats, including the 2009 H1N1 influenza pandemic. In addition to her H1N1-specific work, she serves as a scientific adviser for a CDC cooperative agreement with the Harvard School of Public Health to assess the public's knowledge, attitudes, and behavior in response to health threats. Before joining the ECS Branch, she spent 10 years in CDC's Division of Violence Prevention conducting research on violence-related issues and topics. Lubell received her PhD in sociology from Indiana University, Bloomington, where her dissertation research focused on gender differences in the impact of social isolation and mental health problems on suicide mortality.

Gloria Mark is a professor in the Department of Informatics, University of California, Irvine. Mark received her PhD in psychology from Columbia University. Previously, she was a research scientist at the German National Research Center for Information Technology (GMD), a visiting research scientist at the Boeing Company, and a research scientist at the Electronic Data Systems Center for Advanced Research. Mark's research focuses on the use of technology to support collaboration. Her current projects include studying citizen use of social media for resilience in crises and multitasking in the workplace. A recipient of a Fulbright scholarship, Mark has published more than 100 peer-reviewed publications in the fields of human-computer interaction and computer-supported cooperative work (CSCW). She is the program chair for the Association of Computing Machinery's CSCW'12 and is on the editorial board of the *ACM*

Transactions on Computer-Human Interaction, CSCW Journal, and *Ecommerce Quarterly.*

Dennis Mileti is Professor Emeritus at the University of Colorado, Boulder, where he chaired the Department of Sociology and directed the Natural Hazards Center—the nation's clearinghouse for social science research on hazards and disasters. Mileti is the author of more than 100 publications. Most of these are on the societal aspects of hazards and disasters. His book *Disasters by Design* summarized knowledge in all fields of science and engineering and made recommendations for shifts in national policies and programs. It became the most cited source on natural hazards in the nation and was required reading in more university emergency management courses than any other book in the nation for almost a decade. Mileti has more than four decades of research and applications experience regarding pre-event public preparedness and event-specific disaster warning response; he was awarded the U.S. Army's Civilian Medal of Honor for his work in overseeing investigations by the Army Corps of Engineers about the levee failures during Hurricane Katrina; and he designed the National Institute of Science and Technology's congressional study of evacuation of the World Trade Center towers on 9/11. Mileti has served on many advisory boards, including as chair of the Committee on Disasters in the National Research Council of the National Academies, chair of the Board of Visitors to FEMA's Emergency Management Institute, a board member for the Earthquake Engineering Research Institute, and an advisory council member for the Southern California Earthquake Center. He was appointed by the governor as a California seismic safety commissioner. And he has decades of experience as a consultant to the private and public sectors in matters related to emergency management, including, for example, utilities with nuclear power plants, federal and state agencies, and local governments. Mileti is currently a researcher with a Department of Homeland Security's national center of research excellence on terrorism at the University of Maryland.

Leysia Palen is an associate professor of computer science at the University of Colorado, Boulder, and a faculty fellow with the Institute for the Alliance of Technology, Learning and Society (ATLAS) and the Institute of Cognitive Science. She is the director of the Connectivity Lab and the NSF-funded Project EPIC: Empowering the Public with Information in Crisis. She examines sociotechnical systems, including coordination in online settings as well as the impacts of social computing in off-line arenas and social structures. Her most recent work is in the area of crisis informatics, although she has worked in aviation, digital privacy behavior, personal information management, mobile technology diffusion, health care, and cultural heritage. Prior to her appointment at Colorado, she completed her PhD at the University of California, Irvine, in information and computer science and her undergraduate education in cognitive science at the University of California, San Diego. In 2006, Palen was awarded an NSF Faculty Early Career Development grant for her "Data in Disaster" proposal to study information dissemination in disaster events. From 2005 to 2006, Palen was a visiting professor at the University of Aarhus, Denmark.

Brad Panovich is the chief meteorologist at WCNC-TV in Charlotte, North Carolina. He completed his BS in meteorology at Ohio State University. After OSU, he joined an NBC station in Dayton, Ohio, doing the morning shift. Given an incredible opportunity, Panovich moved to Traverse City, Michigan, to start a new weather department at a Fox affiliate. During the 2005 hurricane season, he reported on both Hurricane Katrina and Hurricane Rita, working with WWL-TV, a CBS affiliate. Panovich shares his weather knowledge on twitter, @wxbrad, and at wxbrad.com.

Paul Resnick is a professor at the University of Michigan School of Information. He previously worked as a researcher at AT&T Labs and AT&T Bell Labs, and as an assistant professor at the MIT Sloan School of Management. He received master's and PhD degrees in electrical engineering and computer science from MIT and a bachelor's degree in mathematics from the University of Michigan. Resnick's research focuses on sociotechnical capital, productive social relationships that are enabled by the ongoing use of information and communication technology. His current projects include making recommender systems resistant to manipulation through rater reputations, nudging people toward politically balanced news consumption and health behavior change, and crowdsourcing fact-correction on the Internet. Resnick was a pioneer in the field of recommender systems (sometimes called collaborative filtering or social filtering). Recommender systems guide people to interesting materials based on recommendations from other people. The GroupLens system he helped develop was awarded the 2010 ACM Software Systems Award. His articles have appeared in *Scientific American, Wired, Communications of the ACM, The American Economic Review, Management Science*, and many other venues. He has a forthcoming MIT Press book (co-authored with Robert Kraut) titled *Building Successful Online Communities: Evidence-based Social Design*.

Dan Roth is a professor in the Department of Computer Science and the Beckman Institute at the University of Illinois at Urbana-Champaign and a University of Illinois Scholar. He is the director of a DHS Center for Multimodal Information Access & Synthesis (MIAS) and also has faculty positions in statistics and in linguistics and at the School of Library and Information Sciences. Roth is a fellow of the ACM and of the AAAI for his contributions to the foundations of machine learning and inference and for developing learning-centered solutions for natural-language-processing problems. He has published broadly in machine learning, natural-language processing, knowledge representation and reasoning, and learning theory and has developed advanced machine-learning-based tools for natural-language applications that are being used widely by the research community. Roth has given keynote talks at major conferences, including the AAAI's Empirical Methods on Natural Language Processing and the European Conference on Machine Learning, and has presented several tutorials in universities and conferences, including at the Association for Computational Linguistics (ACL) and the European Association for Computational Linguistics. Roth was the program chair of AAAI'11, the Conference on Natural Language Learning'02, and ACL'03; is or has been on the editorial board of several journals in his research areas; and has won several teaching and paper awards. Roth received his BA summa cum laude in mathematics from the Technion, Israel, and his PhD in computer science from Harvard University.

W. David Stephenson is principal of Stephenson Strategies (Medfield, Mass.) and an internationally recognized theorist and consultant on the use of mobile devices and social media in preventing and responding to natural disasters and terrorist attacks. Among other accomplishments, he has designed emergency communications strategies for the Wireless Foundation and National Public Radio. He created the first comprehensive guide to terrorism and disasters for smart phones, "The Terrorism Survival Planner." Stephenson is also an international thought leader in the field of open data and the Internet of Things, having written *Data Dynamite: How Liberating Information Will Transform Our World*. He is a graduate of Haverford College and earned an MA from the Newhouse School at Syracuse University.

Peter P. Swire is the C. William O'Neill Professor of Law at the Moritz College of Law of Ohio State University. He is a senior fellow with the Future of Privacy Forum and is also a fellow with the Center for American Progress and Center for Democracy and Technology. He has been a recognized leader in privacy, cybersecurity, and the law of cyberspace for well over a decade, as a scholar, government official, and participant in numerous policy, public interest, and business settings. From 2009 until August 2010 Swire was special assistant to the President for economic policy, serving in the National Economic Council under Lawrence Summers. From 1999 to early 2001 Swire served as the Clinton Administration's chief counselor for privacy, in the U.S. Office of Management and Budget, as the only person to date to have government-wide responsibility for privacy issues. Among his other activities when at OMB, Swire was the White House coordinator for the HIPAA Medical Privacy Rule, led a White House working group on how to update wiretap laws for the Internet age, and led a project on public records and privacy. Swire is lead author of *Information Privacy: Official Reference for the Certified Information Privacy Professional.* Many of his writings appear at www.peterswire.net.

Murray Turoff is a Distinguished Professor Emeritus at the New Jersey Institute of Technology. He is a co-editor of the recent book *Emergency Management Information Systems* (M.E. Sharp, 2010). Besides his early and continuing work with the Delphi method, he spent most of his academic research career in the design and evaluation of computer-mediated communication (CMC) systems. He designed the first collaborative emergency management information system used to manage the wage price freeze in 1971 and many shortages and natural disasters over the following decade. He was with the Office of Emergency Preparedness until 1973. In 1973 he joined NJIT, where he developed the EIES (Electronic Information Exchange System), an operational CMC system for field trials of alternative real user communities which operated until the mid-1990s at NJIT. This effort was sponsored by NSF to support emerging "invisible colleges." A great many communities of practice started on EIES as well as very early work on online learning, including the design and application of the first major virtual classroom system. He is co-author of the book *The Network Nation* with Roxanne Hiltz in 1978 (revised edition 1993, MIT Press), which predicted many features and applications of the current Internet. After 9/11 he turned his attention to his early work in emergency management and related work in planning and foresight and Delphi design. In 2004, he was a cofounder of the international organization ISCRAM (Information Systems for Crisis Response and Management).

Bryan Ware is chief company strategist, lead designer, and primary developer of Digital Sandbox, Inc.'s patented terrorist risk algorithms and is also responsible for designing the breakthrough DS7 product line and its underlying predictive analytics platform. He has been at the forefront of national innovation in risk analytics for more than 15 years, leading threat mitigation technologies for terrorist attacks and helping to identify and overcome military vulnerabilities, protect against bioterrorism, boost infrastructure protection, contend with drug trafficking, and address weapons of mass destruction. A frequent speaker, he is consulted regularly by government and industry executives on security and analytical risk management issues. Ware holds a bachelor of science in applied optics from the Rose-Hulman Institute of Technology.

Duncan Watts is a principal researcher at Microsoft Research and a founding member of the MSR-NYC lab. From 2000 to 2007, he was a professor of sociology at Columbia University, and then, prior to joining Microsoft, a principal research scientist at Yahoo! Research, where he directed the Human Social Dynamics group. He is a former external professor of the Santa Fe Institute and is currently a visiting fellow at Columbia University and at Nuffield College, Oxford. His research on social networks and collective dynamics has appeared in a wide range of journals, from *Nature, Science*, and *Physical Review Letters* to the *American Journal of Sociology* and *Harvard Business Review*. He is also the author of three books: *Small Worlds: The Dynamics of Networks Between Order and Randomness* (Princeton, 1999), *Six Degrees: The Science of a Connected Age* (Norton, 2003), and, most recently, *Everything Is Obvious (Once You Know the Answer)* (Crown Business, 2011). He holds a BSc in physics from the Australian Defence Force Academy, from which he also received his officer's commission in the Royal Australian Navy, and a PhD in theoretical and applied mechanics from Cornell University.

C

Biosketches of Committee and Staff Members

Robert E. Kraut (*Chair*) is the Herbert A. Simon Professor of Human-Computer Interaction at Carnegie Mellon University. He has broad interests in the design and social impact of computing and has conducted empirical research on online communities, the social impact of the Internet, the design of information technology for small-group intellectual work, the communication needs of collaborating scientists, the impact of computer networks on organizations, office automation and employment quality, and technology and home-based employment. His research in specific areas examines in detail the challenges groups currently face in performing social tasks, explores designing new technology to meet some of these challenges, and evaluates the usefulness of the new technology. This cycle of needs assessment, technological design, and evaluation involves both scholarly and applied products. His work on video systems for informal communication, technology for allocating human attention, and online communities follows this model. His recent research has focused on the analysis and design of online communities, such as Usenet groups, guilds in multi-player games, and the editors who write Wikipedia. With collaborators, he is writing *Designing from Theory: Using the Social Sciences as the Basis for Building Online Communities.* He also conducts research on the Internet's role in the interrelationships among firms and on the dynamics of the family. These networks increase the efficiency with which firms can search for or exchange information with each other, but they also shift the type of information that can be easily exchanged, from personal to quantitative. The research examines how these shifts in the cost and quality of communication may influence inter-firm loyalties and market relationships. At the level of the family, the research examines how easy access to remote and personalized information sources and communication partners changes the family's dependence on local resources, among other topics. He wrote a biographical essay, "Re-engineering Social Encounters," in 2003 for the American Psychological Association. In 1980, his research on the evolution of the human face won a Proxmire Golden Fleece award. His biographical essay, "Why Bowlers Smile," and Ed Diener's essay, "Why Robert Kraut Smiles," describe the legacy of that award. Kraut received his BA from Lehigh University in 1968 and his PhD from Yale University in 1973.

Alessandro Acquisti is an associate professor of information technology and public policy at the Heinz College, Carnegie Mellon University. He is the co-director of the CMU Center for Behavioral Decision Research (CBDR), a member of the Carnegie Mellon Cylab, and a fellow of the Ponemon Institute. His work investigates the economic and social impact of information technologies, and in particular the economics and behavioral economics of privacy and information security, as well as privacy in online social networks. His research has been disseminated through journals (including the *Proceedings of the National Academy of Sciences*, *Marketing Science*, *Journal of Consumer Research*, *Marketing Letters*, *Information Systems Research*, *IEEE Security & Privacy*, *Journal of Comparative Economics*, *Rivista di Politica Economica*, and so forth), edited books (*Digital Privacy: Theory, Technologies, and Practices* [Auerbach, 2007]), book chapters, international conferences, and international keynote addresses. His findings have been featured in media outlets such as NPR, NBC, MSNBC.com, the *Washington Post*, the *New York Times* and the *New York Times Magazine*, the *Wall Street Journal*, *New Scientist*, CNN, Fox News, and Bloomberg TV. Acquisti has received national and international awards, including the PET Award for Outstanding Research in Privacy Enhancing Technologies, the IBM Best Academic Privacy Faculty Award, the Heinz College Teaching Excellence Award, and various best paper awards. Two of his manuscripts were selected by the Future of Privacy Forum in their best Privacy Papers for Policy Makers competition. He is and has been a member of the program committees of various international conferences and workshops, including ACM EC, PET, WEIS, ETRICS, WPES, LOCA, QoP, and the Ubicomp Privacy Workshop at Ubicomp. In 2007 he co-chaired the DIMACS Workshop on Information Security Economics and the WEIS Workshop on the Economics of Information Security. In 2008, he co-chaired the first Workshop on Security and Human Behavior with Ross Anderson, Bruce Schneier, and George Loewenstein. His research has been funded by the National Science Foundation, the Humboldt Foundation, the National Aeronautics and Space Administration, and Microsoft Corporation, as well as the CMU CyLab and the CMU Berkman Fund. Prior to joining CMU, Acquisti was a researcher at Xerox PARC in Palo Alto, California, with Bernardo Huberman and the Internet Ecologies Group (as an intern), and for 2 years at RIACS, NASA Ames Research Center, in Mountain View, California, with Maarten Sierhuis and Bill Clancey (as a visiting student). At RIACS, he worked on agent-based simulations of human-robot interaction aboard the International Space Station. While studying at Berkeley, he co-founded with other fellow students a privacy technology company, PGuardian Technologies. In a previous life, Acquisti worked as a classical music producer and label manager (PPMusic.com) and as a freelance arranger, lyrics writer, and soundtrack composer for theatre, television, and indie cinema productions (including works for BMG Ariola/Universal and RAI 3 National Television), and he raced a Yamaha TZ 125 in the USGPRU national championship. Acquisti has lived and studied in Rome (Laurea, economics, University of Rome), Dublin (M.Litt., economics, Trinity College), London (M.Sc., econometrics and mathematical economics, LSE), and in the San Francisco Bay area, where he worked with John Chuang, Doug Tygar, Florian Zettelmeyer, and Hal Varian and received a master's and a PhD in information management and systems from the University of California, Berkeley.

Jon M. Kleinberg is a professor in the Department of Computer Science at Cornell University. His research interests center on algorithmic issues at the interface of networks and information, with an emphasis on the social and information networks that underpin the Web and other online media. He is the recipient of an NSF Career Award, an ONR Young Investigator Award, research fellowships from the MacArthur, Packard, and Sloan foundations, teaching awards from the Cornell Engineering College and Computer Science Department, the Rolf Nevanlinna Prize from the International Mathematical Union, and the National Academy of Sciences Award for Initiatives in Research. Kleinberg received a BS in computer science from Cornell University in 1993 and a PhD, also in computer science, from the Massachusetts Institute of Technology in 1996.

Leslie Luke is the group program manager for the County of San Diego's Office of Emergency Services, where he oversees the Planning Branch, Info/Intel Branch, Recovery Branch, and Support Services. Luke is the recovery coordinator for the County of San Diego and has been the recovery operational area lead for five federally declared disasters and numerous state-declared disasters. He coordinates the Continuity of Community Programs and is a liaison with schools, including child care resource centers; the business sector (leads the ReadySanDiego Business Alliance);

and faith-based initiatives. He oversees the office's public awareness/ public education initiatives, special projects, and the student worker/ internship/volunteer program. Luke has worked for the County of San Diego for 22 years—in the Office of Emergency Services since 2004; before that in the Public Safety Group, a division of the County's Chief Administrative Office; and earlier as an investigator for the County Medical Examiner's Office.

Richard G. Muth, appointed executive director of the Maryland Emergency Management Agency by Governor Martin O'Malley on June 1, 2008, has devoted his entire professional career to safeguarding the lives of Maryland citizens by improving public safety and emergency management practices on the federal, state, and local levels. Muth is a 33-year career and volunteer veteran of the Baltimore County Fire Department. He previously chaired the Governor's Emergency Management Advisory Council, served as a two-term president of the Maryland Emergency Management Association, and was a committee member and subsequent chairman of the State Emergency Response Commission. In 1993, Muth was appointed as director of the Office of Emergency Preparedness in Baltimore County. In 1998, he served as the on-scene coordinator of Maryland resources while battling massive wildfires in Florida and was awarded a Governor's Citation for his efforts. That same year, he was honored by the American Red Cross for establishing new protocols between Baltimore County and the Red Cross. In 1999, he was chosen to chair the Baltimore Metro Council Y2K Contingency Planning Group. In 2003, Muth was appointed by Governor Robert Ehrlich to serve as Baltimore County's director of homeland security and emergency management and oversaw the county's Hazardous Materials Program, advanced tactical rescue, fire department communications, and the Chemical Stockpile Program. He has chaired the U.S. Defense Department's Weapons of Mass Destruction Program's Domestic Preparedness Chemical team and has been recognized for his leadership roles in the aftermath of Hurricane Isabel and as Maryland's emergency resource coordinator following Hurricane Katrina. As MEMA's executive director, Muth oversees a staff of 75 people who work closely with state agencies and Maryland's local jurisdictions, coordinating and planning Maryland's response to any disaster. When a disaster occurs, whether it is man-made or natural, Muth has the primary responsibility for managing the emergency event and closely advising the governor on preparedness and response strategies. Muth holds a certificate in religious studies from St. Mary's Seminary and University, Ecumenical Institute of Theology, in Baltimore.

Leysia Palen is an associate professor of computer science at the University of Colorado, Boulder, and a faculty fellow with the Institute for the Alliance of Technology, Learning and Society (ATLAS) and the Institute of Cognitive Science. She is the director of the Connectivity Lab and the NSF-funded Project EPIC: Empowering the Public with Information in Crisis. She examines sociotechnical systems, including coordination in online settings as well as the impacts of social computing in off-line arenas and social structures. Her most recent work is in the area of crisis informatics, although she has worked in aviation, digital privacy behavior, personal information management, mobile technology diffusion, health care, and cultural heritage. Prior to her appointment at Colorado, she completed her PhD at the University of California, Irvine, in information and computer science and her undergraduate education in cognitive science at the University of California, San Diego. In 2006, Palen was awarded an NSF Faculty Early Career Development grant for her "Data in Disaster" proposal to study information dissemination in disaster events. From 2005 to 2006, Palen was a visiting professor at the University of Aarhus, Denmark.

Timothy L. Sellnow is a professor of communication at the University of Kentucky, where he teaches courses in risk and crisis communication. Sellnow's research focuses on bioterrorism, pre-crisis planning, and communication strategies for crisis management and mitigation. He has conducted funded research for the Department of Homeland Security, the United States Department of Agriculture, and the Centers for Disease Control and Prevention. He has published numerous refereed journal articles on risk and crisis communication and has co-authored four books on risk and crisis communication. His most recent book is *Risk Communication: A Message-Centered Approach*. He is also past editor of the National Communication Association's *Journal of Applied Communication Research*. Sellnow received his PhD from Wayne State University in 1987.

Michele Wood is an assistant professor in the Health Science Department at the California State University, Fullerton, where she teaches courses in statistics and program design and evaluation. She has 20 years of experience designing, implementing, and evaluating interventions. Through her affiliation with the Southern California Injury Prevention Center in the UCLA School of Public Health, she managed a national household preparedness survey conducted as part of the National Center for the Study of Terrorism and Responses to Terrorism (START) Program through the University of Maryland's Center of Excellence, as well as a California household telephone survey of earthquake preparedness. Wood received her PhD in public health from the Department of Community Health

Sciences at the University of California, Los Angeles; she also holds a master's degree in community psychology.

Staff

Jon Eisenberg is the director of the Computer Science and Telecommunications Board of the National Research Council. He has also been the study director for a diverse body of work, including a series of studies exploring Internet and broadband policy and networking and communications technologies. In 1995-1997 he was an American Association for the Advancement of Science (AAAS) Science, Engineering, and Diplomacy Fellow at the U.S. Agency for International Development, where he worked on technology transfer and information and telecommunications policy issues. He received his PhD in physics from the University of Washington in 1996 and a BS in physics with honors from the University of Massachusetts at Amherst in 1988.

Virginia Bacon Talati is an associate program officer for the Computer Science and Telecommunications Board of the National Research Council. She formerly served as a program associate with the Frontiers of Engineering program at the National Academy of Engineering. Prior to her work at the National Academies, she served as a senior project assistant in education technology at the National School Boards Association. She has a BS in science, technology, and culture from the Georgia Institute of Technology and an MPP from George Mason University, with a focus in science and technology policy.

Eric Whitaker is a senior program assistant at the Computer Science and Telecommunications Board of the National Research Council. Prior to joining CSTB, he was a realtor with Long and Foster Real Estate, Inc., in the Washington, D.C., metropolitan area. Before that, he spent several years with the Public Broadcasting Service in Alexandria, Virginia, as an associate in the Corporate Support Department. He has a BA in communication and theater arts from Hampton University.